Biomarkers for Stress in Fish Embryos and Larvae

Biomarkers for Stress in Fish Embryos and Larvae

Irina Rudneva

Ichthyology Department
Institute of Biology
Southern Seas National Ukrainian
Academy of Sciences
Sevastopol
Ukraine

CRC Press
Taylor & Francis Group
Boca Raton London New York

CRC Press is an imprint of the
Taylor & Francis Group, an **informa** business

A SCIENCE PUBLISHERS BOOK

CRC Press
Taylor & Francis Group
6000 Broken Sound Parkway NW, Suite 300
Boca Raton, FL 33487-2742

© 2014 Copyright reserved
CRC Press is an imprint of Taylor & Francis Group, an Informa business

First issued in paperback 2019

No claim to original U.S. Government works

ISBN 13: 978-0-367-45218-6 (pbk)
ISBN 13: 978-1-4822-0738-5 (hbk)

Visit the Taylor & Francis Web site at
http://www.taylorandfrancis.com
CRC Press Web site at
http://www.crcpress.com

To my dear teacher
Professor Dina Alexandrovna Sorkina

Preface

At present many of the aquatic bodies are exposed to the environmental stress factors, which modify the status of water ecosystems and biota. Among aquatic organisms, early developmental stages of fish and invertebrates are the most sensitive to stressors. Sewage and chemical pollution from industrial, agricultural, transport and domestic sectors are the main contribution to the contamination of the sea. They contain the mixture of various kinds of toxicants which damage embryos and larvae resulting in the increase of malformation rate, mortality, abnormalities of development and growth, reproduction process and loss of biodiversity. In addition, knowledge about early-life stages in fish (eggs, larvae, fry and smolts) is very important for the understanding of developmental mechanisms in early life in aquaculture for the optimization of the process of fish hatching, and growth and development in artificial systems. Chemical, physical, and microbial factors cause malformations in fish eggs and larvae, and biomarkers and bioindicators may evaluate fish health in early life. Regarding wild populations, knowledge about fish early life may provide insight into how various fish species adapt to their environment and survive under various conditions including anthropogenic impact. The aim of the present study is to show the role of bioindicators and biomarkers in fish early life in normal and stress conditions, the formation of defense systems during early development and their response to damage factors both chemical and physical. The present study demonstrates the important role of the multi-biomarker approach in the detection of biological effects of pollution in fish early life and their applications in evaluation of water quality and risk assessment.

Contents

Introduction

Monitoring of fish development in early life is very important for understanding of the mechanisms of gene expression and interaction between exogenous and endogenous factors. Undoubtedly the development of fish in early life depends on the specificity of their ecology and biology and how they adapt to habitats. The early development of various fish species depends on both exogenous (temperature, salinity, oxygen consumption, pressure, feeding, etc.) and endogenous factors (egg and embryo size, growth, time of embryonic development, biochemical composition of yolk sac, etc.). The type of feeding also plays a role in fish embryo development because endogenous/exogenous feeding distinguishes the developmental phases of embryo (egg), eleutheroembryo (feeding off the yolk sac) and larvae (exogenous feeding) in fish.

In addition, the development of fish embryo changes under the impact of unfavorable physical, chemical and microbial factors and the toxic effects of these may destroy the process of normal development because at their early developmental stages fish are very sensitive to environmental fluctuations both natural and anthropogenic. Xenobiotics accumulation, anoxia and hypoxia have a negative influence on early development and cause stress in embryo and larvae. Several investigators have documented the responses of fish embryos to pollution and larvae to environmental stress leading to increase in the malformation rate, high mortality, changes in physiological and biochemical status which in turn have resulted in reduced growth, reproductive impairment and loss of biodiversity. Thus it is important to make a detailed study of a fish's early life and to know the critical periods for applying this information to aquaculture and for understanding the developmental mechanisms, adaptive strategy of fish in its early growth phase and how it impacts hatching process and larvae survival.

The effects of contamination on physiological and biochemical processes lead to different kinds of adverse changes. Thus the measurements of behavior, feeding, growth rate, malformation rate, respiration, locomotion, osmoregulation, pulse allow for evaluation of toxic effects of xenobiotics or environmental toxicity. In conditions of acute pollution and the exposure to high doses of pollutants the physiological processes are damaged and often lead to death of the organism. Fish embryonic tests are applicable for measuring different kinds of toxicant classification and labeling under experimental conditions. The use of fish embryo toxicity test (FET) data in various types of environmental assessments is a good tool for evaluation of hazards, risk, effluent and ecological status of water bodies. Fish display a wide range of developmental ontogenesis depending on their taxonomic, evolutionary, and ecological situation which is important to know besides knowing the practical implications on the use of aquatic toxicity test on early fish development stages. Thus developing embryos and larvae are applied as biomonitors for evaluation of ecological status of polluted and non-polluted marine areas.

However, many toxicologists have studied the survival (or mortality) of the test-organisms which are used for toxicity determination; death as endpoint of the experiment is considered an insensitive bioindicator. Such physiological responses give scant information about environmental pollution and its impact on the organism. Physiological and biochemical assays are useful for indicating the "early warning" biomarkers which appear at the beginning of stress reaction. Biomarkers are very sensitive to very low stressors concentrations and their fluctuations should be monitored and compared with the normal ranges. At present many methods are used for biomarker determination such as physiological, biochemical, and genetic. The relationship between pollutant exposure and response to biomarkers of fish embryos and larvae is well documented. But the trends of the biomarkers fluctuations in toxicants-treated organisms are varied widely and in some cases the authors obtained contradictory results.

The evaluation of toxicity in a sensitive fish ontogenic stages (embryos, larvae, fry and smolts) should constitute an early warning of population decline and hence an appropriate and ecologically relevant endpoint. Because fish early developmental stages are very sensitive to anthropogenic pollution, some of them may be tested as biomonitors for the evaluation of the ecological status and risk assessment of aquatic environments. Fish biochemical parameters could be directly related to their habitats. Thus the additional information of biomarkers application for monitoring purposes in different aquatic areas could be helpful for development of their use in the evaluation of environment ecological status.

Hence, in the present study we have tried to apply the biomarker approach of fish early life stages and to demonstrate the development of biomarker systems in fish embryogenesis, their functions in defense against unfavorable factors in experimental and field conditions. Taking into account the growing interest in aquaculture research, the increase in tools and methods of the evaluation of fish early development also is very important. Fish embryos and larvae combine a number of key experimental advantages because they develop rapidly and they may incubate in artificial systems and get exposed to various toxicants in laboratory conditions. We hope that the data obtained will be helpful for the researchers in aquatic sciences, aquaculture, ecology, ecotoxicology and ontogenesis.

The author would like to thank all colleagues and personally Dr. Lydia. S. Oven, Dr. Nadejda F. Shevchenko, Dr. Tatyana L. Chesalina, Dr. Natalia. S. Kuzminova, Dr. Tatyana. N Klimova and Valentin G. Shaida for larvae collection, developmental stages and species taxa determination and for participation in toxicological experiments.

CHAPTER 1

Stress Response in Fish

Stress is the nonspecific response of the organism to any demand made upon it (Selye, 1973). According to Selye (1984) stress should be divided into two phases: "eustress" or the healthy stress and "distress" or the bad stress. Eustress occurs as a response of the organism undergoing situations that provoke physiological changes that optimize its biological performance and defense mechanisms. Distress occurs when certain factors promote physiological changes in an organism that may compromise the organism's integrity. Major part of stress research is focused on distress phase (Martinez-Porchas et al., 2009).

The environmental factor that elicits stress is "stressor". According to the Selye concept of stress, the living organisms are constantly compensating for effects of environmental stressors suggesting that normal physiological conditions change over time and are dependent on individual and environmental characteristics. The magnitude of stress that can be tolerated by an organism is not dependent on whether the source of the stressor is natural or anthropogenic; rather, it is a function of how well the animal is equipped to compensate for the effects of exposure. Compensatory ability of an organism in a given environment is dependent on its evolutionary history and the reserve energy that can be allocated to offset effects of a new stressor. Cumulative effects of multiple stressors can deplete an organism's reserves, thereby reducing its ability to cope. Conversely, organisms in relatively benign environments with abundant food may have large energy reserves that can be mobilized to maintain homeostasis. The environmental background that an organism lives in plays an important role in its ability to compensate for effects of contaminant exposure. Stress is defined as a condition that disturbs the normal functioning of the biological system or a condition that decreases fitness. Stress can be extrinsic (environmental) or intrinsic stress factors such as genetic stress (e.g., inbreeding and deleterious mutations) and aging (SØrensen et al., 2003).

Stress modifies physiological and biochemical status of the organisms and it causes various kinds of disturbances. Common stressors encountered by captive fish include physical and mental trauma associated with capture, transport, handling, and crowding; malnutrition; variations in water temperature, oxygen, and salinity; and peripheral effects of contaminant exposure or infectious disease. Some stress responses are detectable through gross or microscopic examination of various organs or tissues. Stress responses are most consistently observed in the gills, liver, skin, and components of the urogenital tract (Harperand Wolf, 2009). The measurements of behavior, respiration, locomotion, osmoregulation, pulse, biochemical parameters (biomarkers) changes are help in the evaluation of toxic effects of xenobiotics or environmental toxicity. In acute pollution and exposure to high doses of pollutants the physiological processes are damaged and often lead to death of the organism. However, in many toxicological studies the mortality of the test-organisms are used for toxicity evaluation and death as endpoint of the experiment is applied as a bioindicator. Such physiological responses give little information about environmental pollution and its impact on the organism. Physiological and biochemical assays are useful for indicating the "early warning" biomarkers which appear at the beginning of stress reaction. Such biomarkers even show up at very low stressor concentrations and their fluctuations can be monitored and compared with the normal ranges (Adams, 1994, 2005; Adams et al., 1992a,b).

Stress response in teleost fish shows many similarities to that of the terrestrial vertebrates. These concern the principal messengers of the brain-sympathetic-chromaffin cell axis (equivalent of the brain-sympathetic-adrenal medulla axis) and the brain-pituitary-interrenal axis (equivalent of the brain-pituitary-adrenal axis), as well as their functions, involving stimulation of oxygen uptake and transfer, mobilization of energy substrates, reallocation of energy away from growth and reproduction, and mainly suppressive effects on immune functions. There is also growing evidence of intensive interaction between the neuroendocrine system and the immune system in fish. However, conspicuous differences are present, and these are primarily related to the aquatic environment of fish. In addition, among almost 20,000 known teleost species, there are many indications that the stress response is variable and flexible in fish, in line with the great diversity of adaptations that enable these animals to live in a large variety of aquatic habitats (Bonga, 1997).

The response to stress in fish is considered an adaptive mechanism that allows the fish to cope with real or perceived stressors in order to maintain its normal or homeostatic state. If the intensity of the stressor is overly severe or long-lasting, physiological response mechanisms may be compromised and can become detrimental to the fish's health and well-being, or maladaptive, a state associated with the term "distress" (Barton, 2002). The response to stress in fish is characterized by the stimulation of the hypothalamus, which results in the activation of the neuroendocrine system and a subsequent cascade of metabolic and physiological changes. These changes enhance the tolerance of an organism to face an environmental variation or an adverse situation while maintaining a homeostasis status (Martinez-Porchas et al., 2009).

In fish stressors increase the permeability of the surface epithelia, including the gills, to water and ions, and thus induce systemic hydromineral disturbances. High circulating catecholamine levels as well as structural damage to the gills and perhaps the skin are prime causal factors. This is associated with increased cellular turnover in these organs. In fish, cortisol combines glucocorticoid and mineralocorticoid actions, with the latter being essential for the restoration of hydromineral homeostasis, in concert with hormones such as prolactin (in freshwater) and growth hormone (in seawater) (Barton, 2002).

Fish are exposed to aquatic pollutants via the extensive and delicate respiratory surface of the gills and, in seawater, also via drinking. The high bioavailability of many chemicals in water is an additional factor. Together with the variety of highly sensitive perceptive mechanisms in the integument, this may explain why so many pollutants evoke an integrated stress response in fish in addition to their toxic effects at the cell and tissue levels. Exposure to chemicals may also directly compromise the stress response by interfering with specific neuroendocrine control mechanisms. Because hydromineral disturbance is inherent to stress in fish, external factors such as water pH, mineral composition, and ionic calcium levels have a significant impact on stressor intensity (Barton, 2002).

Development of stress response of the organism involves three sequential phases (Beyers et al., 1999):

The first phase of the general adaptation syndrome is the stage of physiological alarm. During this stage, the effects of stressors (chemical or physical exposure) upset homeostasis of the organism. As the physiological system adjusts to compensate for specific effects from the mode of action of the contaminant, a host of nonspecific homeostatic mechanisms are also induced in order to reestablish equilibrium. In fish, this stage may be associated with a loss of appetite, loss of equilibrium, locomotion and behavioral changes. The advantage of the behavioral response is due to the integration of biochemical and physiological processes that reflect at higher levels of organization with ecological relevance. This rapid response would reflect a defense mechanism protective against further exposure and potential development of more pronounced deleterious relocation of a species with negative consequences (Hellou, 2011). Sometimes fish leave the impacted habitats and migrate to unstressed areas.

The second phase of the general adaptation syndrome is the stage of resistance. This stage occurs when physiological adaptation to the stressor has been accomplished. During this stage, compensating for the effects of a chemical stressor becomes part of the normal cost of living for an exposed animal. This stage may be associated with increased metabolic rate. Most organisms are not adapted through evolution to the energetic costs of compensating for anthropogenic stressors. Contaminant effects can be compensated, but if the stressor is of sufficient magnitude and is applied for a sufficient length of time.

The third phase of the general adaptation syndrome is the stage of exhaustion. During this stage, cumulative effects of long-term exposure to sublethal stressor result in premature death of the individual. In the case of chemical stressor, mortality occurs because the physiological system responsible for compensating for toxic effects becomes exhausted and stops functioning. Exhaustion occurs because the physiological system has been forced to function at a faster rate than that for which it is evolutionarily adapted.

Fish respond to a stressor by eliciting a generalized physiological stress response, which is characterized by the corresponding changes in various levels of their biological organization. For the first time stress hormones increase and consequent changes help maintain the organism's normal or homeostatic state (Iwama et al., 2004). The response to a stress is a dynamic process and physiological measurements taken during a time period are only representative instantaneous "snap-shots" of the process. A significant delay, depending on the level and type of response, can occur from initial perception of the stressor by CNS to the time when plasma cortisol or other features of interest reach a peak level of response. This response has been considered to be adaptive and it includes increases in plasma cortisol, catecholamines and glucose levels, increases in branchial blood flow and increase in muscular activity. However, pathological states, even death, can result if the magnitude or duration of the stressor overwhelms the animal's ability to compensate for the negative influences of the stressor (Barton, 2002).

The effect of environmental contaminants on health is a major concern because exposure is associated with a number of diseases, including cancer, inflammation, microbial and virus infections. Developing organism is extremely sensitive to toxic effects, because it absorbs more pollutants compared to its weight than an adult. It has become increasingly clear that some cancers and birth defects stem from common exposures that occur early in life or even before conception. Paternal and maternal exposure can induce germ cell cancers in larvae. Male organisms are at higher risk than female for a number of congenital abnormalities (Scherb and Voigt, 2011).

Stressors impact the total organism, tissues, organs and cells. Animals reply to stressors in various levels on their biological organization - from the total organism to molecules (Iwama et al., 1999). Fish are exposed to stressors both in environmental and in artificial conditions. The detection of fish response on various stressors both specific and nonspecific is very important for understanding of adaptive mechanisms in field and in aquaculture conditions.

1.1 Stressors

The potential stressors are grouped as being environmental, physical or biological. Many stressors are unique to certain species or geographic areas. The stress response of fish has a genetic component, and thus the responses differ from species to species; different stocks or races of the same species differ in their tolerance to applied stressors (Iwama et al., 1999).

When individuals in a population are exposed to stressful conditions, generally three possibilities exist (SØrensen et al., 2003):

- the animals in that population try to avoid the stress, either by moving to a more favorable habitat, by adjusting their activity patterns or by changing into physiological state that might be more resistant (e.g., by going into hibernation or diapause);
- the population can adapt to the stressful condition through selection, or individuals can respond through a plastic response;
- foil in the above and go extinct.

Stressful conditions can thus act as evolutionary forces that populations respond to adaptively. Many environmental factors are likely to affect the distribution and abundance of species and populations in nature (see Chapter 5). Due to anthropogenic activity, e.g., global warming, climate change and pollution, environmental changes might even be more drastic and unpredictable in the future than at present. Habitat fragmentation leading to isolation and reduction of population size are increasing the degree of inbreeding and genetic drift. However, even without human-activity caused interventions, environmental and genetic factors are changing and will affect the ecology and evolution of species (SØrensen et al., 2003).

Environmental stressors include physical and chemical conditions of the water and sediments such as dissolved oxygen, temperature, pH, biogens (nutrients and ammonia) concentration, salinity, hardness, gas content and partial pressures, time of day, wave length of light and its intensity, nutritional factors as well as toxicants both non-organic (heavy metals) and organic (pesticides, PCB, phenols, PAH, etc.) (Fig. 1.1). Domestic, industrial and agricultural waste waters contain different kinds of stressors Phihysical stressors include those that involve handling, crowding, confinement, transport and electromagnetic fields. Commercial fishing and angling also cause stress to fish. Disease pathogens, viruses and parasites can also be considered as biological stressors.

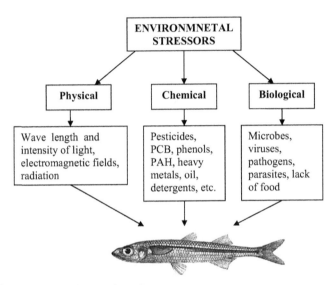

Fig 1.1. Stressors groups in aquatic environments.

Various aspects of intensive aquaculture also stimulate stress consequences in fish because fish in artificial systems are frequently exposed to a range of "unnatural" stressors which are related to rearing practices. These include overcrowding, handling and transport. Activation of the HPI axis is one of the major endocrine responses to such stressors and administration of cortisol has often been used to simulate stress situations and further investigate the physiological consequences of HPI activation (Prunet et al., 2008).

Social subordination associated with the competition for limited resources, such as food and foraging sites is also stressful and characterised by elevated levels of plasma cortisol and brain serotonin metabolism together with other indicators of sustained stress. In the wild, social status is a major determinant of life history traits, reproductive success and survival in many species. In aquaculture regimes, although social hierarchies become less distinct with increasing stocking densities, differences in growth rate are a common phenomenon, and fish that grow poorly may do so because they are socially subordinate. Hence, stress in fish causes abiotic and/or biotic factors and their combination is characterized as following (Barton, 2002):

- differences in behavior, hormone levels and neurochemistry associated with social status in various teleosts especially in aquaculture conditions, appear mainly as a result of chronic stress in subordinates;
- alterations in almost all factors related to physical characteristics of the aquatic environment (temperature, salinity, turbidity, pH, oxygen level, xenobiotics) can result in an integrated stress response in fish if the changes are sudden or sufficiently extreme;
- because they are poikilotherms, fluctuations in environmental temperature cause seasonal stress in fish and induce rapid physiological responses to these changes;
- hypoxia is an important stress factor for all fish species and was among the first stressors to be analyzed using a genomic approach;
- alterations in salinity concentration are stressors which modify homeostasis and the osmoregulatory mechanisms;
- water and sediment pollution arising from anthropogenic activities can have a significant effect on fish physiological status causes integrated "non-specific" stress response. The high sensitivity to environmental pollution that is exhibited by fish is a characteristic of aquatic animals and is related to the intimate contact of respiratory surfaces with the medium together, with the large variety of perceptive mechanisms present in the fish integument (including light, mechano-, temperature, electro-, and chemo-receptors). Thus, exposure to water-borne contaminants leads to an increase in stress-related hormone levels and also results in damage at the level of the gill and skin (Barton, 2002).

Acclimation is traditionally a process occurring over long periods of time (days or weeks) in laboratory conditions. Temperature, or generally stress exposures used, are normally within the threshold for development and can be applied to any life stage or throughout development (developmental acclimation). When the same process occurs in nature it is known as acclimatization. However, this term has also been used interchangeably with acclimation. Hardening is typically a much shorter process (or treatment) to a more extreme, but non-lethal stress condition. The changes brought about by hardening are primarily reversible, whereas acclimation and especially developmental acclimation leads to irreversible changes/hardening and acclimation is known to affect the composition of membrane lipids, energy reserves and (especially for hardening) initiate the stress response including the expression of heat shock proteins (SØrensen et al., 2003). Physiological changes in turn affect many life history and fitness traits such as fecundity, longevity and stress resistance. Physiological changes in stress resistance and life history traits caused by hardening or acclimation have been in the direction predicted from an adaptive hypothesis and acclimation and hardening are generally considered to be adaptive. This expectation is not always met when acclimation is considered. The "beneficial" acclimation hypothesis, which predicts that an organism will perform best under the conditions in which it has been raised (or acclimated) has been the subject of much debate. Probably, acclimation (as the case of hardening) does not optimize performance at the acclimation condition per se, but increases performance to future extremes, as benefits particularly seem exist at the extreme ends of environmental regimes (SØrensen et al., 2003).

Pathohysiological stressors encompass a wide range of potential stressors including those that disturb the fish physically and psychologically. Handling, crowding, confinement, transport, or other forms of physical disturbance to fish also have psychological components. Many of these are practised in the intensive culture of fish for both wild stock enhancement, and for the commercial production of food. Chasing fish to exhaustion, or holding them in a net out of water for 30–60 sec have been common protocols utilized to study acute stress responses in fish. Angling also stresses fish in this manner. Other psychological stressors can manifest in dominance hierarchies, which develop between individuals within confines such as experimental tanks or possibly in natural environments (SØrensen et al., 2003).

Pathogens and parasites can also be considered as biological stressors. Fish disease, and outbreaks leading to massive mortalities, occur in nature as well as in culture stocks. Microalgae blooms and eutrophication can stress fish and kill them in the wild and in aquaculture conditions. Microalgae may kill fish directly by their toxins (siguatera), irritate or severely damage the gill epithelium with their spines, or kill the fish indirectly by hypoxic and anoxic conditions by either lowering water oxygen levels directly or by increasing the diffusion distance between blood and water through the stimulation of mucus production on the gill surface.

Stressful factors can actually exacerbate and attenuate the cortisol response to a second stressor. Continual interregnal activity will down-regulate the HPI axis as a result of negative feedback by cortisol, which causes the attenuation of the response to additional stressors. Cortisol is the major steroid stress hormone in salmonid fish which has diverse effect on several tissues including the myocardium. Severe stress is clearly associated with a poor prognosis in individuals with established cardiac pathology and disease (Johansen et al., 2011). Thus, when a second stressor subsequently challenges fish exposed to a chronic stressor, the corticosteroid response to the additional stressor may be reduced considerably depending upon controls (Barton, 2002).

Fish can exhibit a cumulative response to repeated stressors and the peak cortisol responses after the final disturbance are cumulated. Repeated exposure to mild stressors can desensitize fish and attenuate the neuroendocrine and metabolic responses to subsequent exposure to stressors. The length of time between discrete stressors, the effect of multiple stressors, and the severity of continuous stressors are important factors that will likely influence how fish respond. Unless stressors, singularly or in combination, are severe enough to challenge the fish's homeostatic mechanisms beyond their compensatory limits or permanently alter them, which ultimately may cause death, physiological processes generally adapt to compensate for the stress. In these cases, blood chemistry features, such as cortisol, used to evaluate stress may appear "normal" and alternative approaches, such as determining the magnitude of response to an additional acute stressor, may be needed to assess the fish's physiological status (Barton, 2002; Feldhaus et al., 2008).

1.2 Physiological Stress Response

Physiological responses of fish to stressors have been broadly categorized into the primary, secondary and tertiary responses. The initial *primary response* represents the recognition of a real (physical or chemical) or perceived (presence of predators) stressor by the central nervous system (Barton, 2002). Most stressors include a neuroendocrine/endocrine response, which is characterized by a rapid release of stress hormones (catecholamines (CATS) and cortisol) into the circulation. CATS are released from the chromaffin tissue in the head kidney and also from the endings of adrenergic nerves. Because catecholamines, predominantly epinephrine in teleosts are stored in the chromaffin cells, their release is rapid and the circulation levels of these hormones increase immediately with stress (Barton, 2002).

The release of cortisol in fish is delayed relatively to catecholamine release. The pathway for cortisol release begins in the HPI axis with the release of corticotrophin-releasing hormone (CRH) or factor (CRF) chiefly from the hypothalamus in the brain, which stimulates the corticotrophic cells of the anterior pituitary to secrete adrenocorticotropin (ACTH). Circulating ACTH, in turn, stimulates the interregnal cells (adrenal cortex homologue) embedded in the kidney to synthesize and release corticosteroids into circulation for distribution to target tissues. The interregnal tissue is located in the anterior kidney in teleosts and exhibits considerable morphological variations among taxonomic groups. The results suggest a trend towards lower stress responses in chondrosteans fish compared with teleosts. Among teleosts the level of cortisol response on stress is also varied widely.

Cortisol synthesis and release from interregnal cells have a lag time of several minutes, unlike chromaffin cells. The circulating level of cortisol is commonly used as an indicator of the degree of stress experienced by fish (Barton, 2002).

Most of the fish tested show their highest plasma increase in cortisol within about 0.5–1 hr after stressful disturbance, but some fish species respond after about 4 hr and more. Differences in corticosteroid stress responses also exist among strains or stocks within the same species and between wild and hatchery fish. Long term exposure to cortisol can suppress the immune system, reduce reproductive capacity, decrease growth rates, inhibit smoltification in salmon, and a variety of other actions that are considered detrimental to survival (Barton, 2002).

Cortisol is released from the interrenal tissue, located in the head kidney, in response to several pituitary hormones but most potently to the adrenocorticotrophic hormone (ACTH) which may also stimulate the release of the CAT epinephrine, and chronically increased levels of cortisol may affect CAT storage and release in trout. The physiological action of both hormones are dependent on sensitivity of target tissues. Elevations in plasma cortisol range at least as much as two orders of magnitude among fishes following an identical stressor and can be much higher, depending on species (Barton, 2002).

Other peripheral hormones can become elevated during stress, including notably thyroxine, prolactin, and somatolactine. Also, stress may suppress reproductive hormones in circulation as does an elevated level of cortisol. However, these other hormones have not yet been demonstrated to be useful stress indicators. Characteristic cortisol elevations of fishes in response to acute stressors tend to range within about 30 and 300 ng/ml (Barton, 2002). The resting and stressed levels of adrenaline and cortisol concentrations in plasma of salmonids are: adrenaline < 3 and 20–70 n mole l^{-1} and cortisol < 10 and 40–200 ng ml^{-1} respectively. These values should serve only as general guidelines because individual conditions, including genetic characteristics, prior rearing history and local environment can affect the true plasma values for control and stressed states (Iwama et al., 1998).

The developmental stage of the fish can also affect its responsiveness to a stressor. A fish's ability to respond to a disturbance develops very early in life. Fish larvae showed elevations of whole-body cortisol after they were exposed to different stressors. High cortisol content in embryos and larvae may actually affect larval quality (Barton, 2002).

The *secondary response* comprises the wide range of changes that are caused, to a large extent, by the stress hormones in blood, organs and tissues of the animal. Adrenalin and cortisol activate a number of metabolic pathways that result in alterations in blood chemistry and haemotology and induction of stress proteins synthesis. Stress is an energy demanding process and the animal mobilizes energy substrates to cope with stress metabolically (Iwama et al., 1999). The production of glucose with stress assists the animal by providing energy substrates to tissues such as brain, gills, and muscles, in order to cope with the increased energy demand. The stress hormones adrenaline and cortisol have been shown to increase glucose production in fish, elevation of glycogenolysis level, and likely play an important role in the stress-associated increase in plasma glucose of fish, including nutritional state, can affect response and glucose clearance rates (Adams, 1994; Iwama et al., 1999; Barton, 2002). The measurement of stress hormones together with plasma glucose concentration has been used as an indicator of stress states in fish. It is probably the most commonly measured secondary change that occurs during the stress response in fish (Iwama et al., 1998). For response measurements the biomarkers are used. They are defined as sub-lethal biological measures of stress response to, and effects of pollutants in living organisms. Biomarkers have been identified as a powerful and cost-effective approach to obtain information on state of the environment and the effects of pollution on living biological resources. Generally biomarkers are selected for their responses to particular classes of environmental contaminants rather than to specific individual chemicals, consequently they are sufficiently versatile to enable use in a wide variety of field situations (Goksoyr et al., 1996).

The *tertiary response* represents whole-animal and population level changes associated with stress. In case of inability of fish to acclimate or adapt to the stress, whole-animal changes may occur as a result of energy repartitioning, by diverting energy substrates to cope with the increased energy demand associated with stress. Thus, chronic exposure to stressors, depending on the intensity and duration, can lead to decrease in growth, disease resistance, reproductive success, smolting, swimming performance, malformations and abnormalities of development and other characteristics of the whole animal or population. At a population level, decreased recruitment and productivity may alter community species abundance and lack of biodiversity (Iwama et al., 1998, 1999, 2004).

1.3 Cellular Stress Response

The cell responds to stressors. A dominant aspect of the changes in protein profile as a part of the cellular stress response is the change in the concentration of different classes of stress proteins (see Chapter 2). Liver, kidney and gill tissues seem to be sensitive tissues to the cell response especially Hsps, and they have potential to be used as biomarkers of environmental stress in fish. Hsps play an important role in the cell's response to a wide range of damaging (stressful) conditions and are important to recovery and survival of organisms (SØrensen et al., 2003). The potential for using Hsp70 as a biomarker of cellular stress, and consequently, the increase in tissue concentrations of these proteins may be indicative of physiological stress. However, the relationship between the cellular Hsps response and the physiological stress response appears to be a complex one that requires further study (Iwama et al., 1998).

At present transcriptomic analysis is used for stress response of fish determination and identification of a complex network of genes whose expression is modified after exposure to handling or confinement stress. The alterations in functional classes such as binding and transport of metal ions, chaperone and heat shock proteins, cytoskeleton and microtubules and a number of signaling pathways in the brain were noted. Exposure to an acute and chronic stressors also led to an increase in haptoglobin expression in trout liver and head-kidney, a prominent acute phase protein which has never been previously associated in fish with non-immunological stressors (Prunet et al., 2008).

The potential of oxygen free radicals and other reactive oxygen species (ROS) to damage tissues and cellular components, called oxidative stress, in biological systems plays an important role in stress development. Oxidative stress is also associated with cell response on ROS. All respiring cells produce free radicals in chloroplasts, mitochondria, endoplasmic reticulum, peroxisomes and glyxysomes. ROS may be harmful for biological systems because they cause oxidative stress, and the generation and accumulation of ROS which damage cell membranes, lipids, proteins, and DNA. On the other hand ROS can also act in signal transduction and participate in the formation of intermediates (prostaglandins, leikotriens, etc.); these play an important role for the expression of several transcription factors (heat shock-inducing factor, nuclear factor, the cell-gene p53, nitrogen-activated protein kinase, etc.). Oxidative stress also plays a role in apoptosis in two pathways, the death-receptor and the mitochondrial (Winston, 1990, 1991; Winston and Di Giulio, 1991; Livingstone, 2001; Burlakova, 2005; Lesser, 2006; Vladimirov and Proskurina, 2007). Redox changes in the cell might serve as a regulator of Hsp production (Iwama et al., 1998).

The balance between prooxidant endogenous and exogenous factors (i.e., environmental pollutants) and antioxidant defenses (enzymatic and nonenzymatic low molecular weight antioxidants) in biological systems can be used to assess toxic effects under stressful environmental conditions, especially oxidative damage induced by different classes of chemical pollutants and hypoxia. A higher number of agricultural and industrial chemicals are entering the aquatic environment and are accumulating in tissues of aquatic organisms. Heavy metals, polycyclic aromatic hydrocarbons (PAH), organochlorine and organophosphate pesticides, polychlorinated biphenyls (PCB), dioxins, and other xenobiotics play important roles in the mechanistic aspects of oxidative damage (Valavanidis et al., 2006; Slaninova et al., 2009).

Cellular response to protein oxidation includes two processes: reparation of oxidative damaged proteins and/or degradation of oxidized proteins. After oxidation of proteins, living systems try to rescue defect polypeptides and restore their function. Various shock or stress proteins, both the constitutive and inducible forms, are able to assist reconstitution of tertiary protein structure. Chaperones and special adaptive enzymes are involved in this response. Oxidized/cross-linked material are accumulated in cells and are degraded by special enzymes also (Grune, 2000). The ratio of oxidative products and antioxidants is the important feature of the organism health and its change under unfavorable conditions may be used as a good tool for the evaluation of stressful effects—chemical, physical and biological.

However, several researchers note that stress in fish is an amorphous term that does not have a consistently applied definition; procedures used to determine or measure stress can be inherently stressful; interactions between stressors and stress responses are highly complex; and morphologically, stress responses are often difficult to distinguish from tissue damage or compensatory adaptations induced specifically by the stressor (Harper and Wolf, 2009). Therefore, further investigations are necessary to more precisely define the role of stress in the interpretation of fish research results.

There are many possible applications of measuring the stress response in fish, and other aquatic animals for monitoring purposes, for the evaluation of water quality in aquatic environment and fish physiological status, the optimization of aquaculture conditions, for development of remedies of polluted areas and waters. A battery of various biomarkers and bioindicators of stress is used for the detection of fish health and the status of their habitats. Although many studies have been devoted to understand the effects of pollutants on various physiological and biochemical parameters in fish, in many cases their precise mechanisms of action remain to be clarified because it is important for understanding the mechanisms of fish adaptation in field and aquaculture conditions. However, fish response to unfavorable factors range widely and further investigations are necessary to more precisely define the role of stress in the interpretation of fish research results.

References

Adams, S.M. 1994. Early signs of environmental damage and recovery. Oak Ridge National Laboratory Review. 3: 45–56.

Adams, S.M. 2005. Assessing cause and effect of multiple stressors on marine systems. Mar. Pollut. Bul. 51: 649–657.

Adams, S.M., W.D. Crumby, M.S. Greely, J.B. Jimenez, M.G. Rayn and E.M. Shilling. 1992a. Relationships between physiological and fish population responses in a contaminated stream. Environ. Toxicol. Chem. 11: 1549–1557.

Adams, S.M., W.D. Crumby, M.S. Greely, L.R. Shugart and C.F. Saylor. 1992b. Responses of fish populations and contaminates to pulp mill effluents: a holistic assessment. Ecotoxicol. Environ. Safety. 24: 347–360.

Barton, B.A. 2002. Stress in fishes: a diversity of responses with particular reference to changes in circulating corticosteroids. Integ. Comp. Biol. 42: 517–525.

Beyers, D.W., J.A. Ric, W.H. Clements and C.J. Henry. 1999. Estimating physiological cost of chemical exposure: integrating, energetics and stress to quantify toxic effects in fish. Can J. Fish. Aquat. Sci. 56: 814–822.

Bonga, S.E.W. 1997. The stress response in fish. Physiol. Rev. 77(3): 591–625.

Burlakova, E.B. 2005. Bioantioxidants: yesterday, today, tomorrow. *In*: Chemical and Biological kynetics. New approach. Moscow. Chemistry Publ. Vol. 2, pp. 10–45 (*in Russian*).

Feldhaus, J.W., S.A. Heppell, M.G. Mesa and H.W. Li. 2008. Hepatic Heat Shock Protein 70 and Plasma Cortisol Levels in Rainbow Trout after Tagging with a Passive Integrated Transponder. Transactions of the American Fisheries Society. 137: 690–695.

Goksoyr, A., J. Beyer, E. Egaas, B.E. Grosvik, K. Hylland, M. Sandvik and J.U. Skaare. 1996. Biomarker responses in flounder (*Platichthys flesus*) and their use in pollution monitoring. Mar. Pollut. Bull. 33: 36–45.

Grune, T. 2000. Oxidative stress, aging and the proteasomal system. Biogerontology. 1: 31–40.

Harper, C. and J.C. Wolf. 2009. Morphologic Effects of the Stress Response in Fish. ILAR Journal 50(4): 387–396.

Hellou J. 2011. Behavioral ecotoxicology, an "early warning" signal to assess environmental quality. Environ. Sci. Pollut. Res. 18: 1–11.

Iwama, G.K., P.T. Thomas, R.B. Forsyth and M.M. Vijayan. 1998. Heat shock protein expression in fish. Rev. Fish Biology and Fisheries. 8: 35–56.

Iwama, G.K., M.M. Vuayan, R.B. Forsyth and P.A. Ackerman. 1999. Heat shock proteins and physiological stress in fish. Amer. Zool. 39: 901–909.

Iwama, G.K., L.O.B. Afonso, A. Todgham, P. Ackerman and K. Nakano. 2004. Are Hsps suitable for indicating stressed states in fish? J. Experim. Biol. 207: 15–19.

Johansen, I.B., I.G. Lunde, H. Rosjo, G. Christensen, G.E. Nilsson, M. Bakken and Ø. Øverli 2011. Cortisol response to stress is associated with myocardial remodeling in salmonid fishes. J. Experim. Biol. 214: 1313–1321.

Lesser, M.P. 2006. Oxidative stress in marine environments: biochemistry and physiological ecology. Ann. Rev. Physiol. 68: 253–278.

Livingstone, D.R. 2001. Contaminant-stimulated reactive oxygen species production and oxidative damage in aquatic organisms. Mar. Pollut. Bull. 42: 656–665.

Martinez-Porchas, M., L.R. Martinez-Cordova and R. Ramos-Enriquez. 2009. Cortisol and Glucose: Reliable indicators of fish stress? Pan-American Journal of Aquatic Sciences. 4(2): 158–178.

Prunet, P., M.T. Cairns, S. Winberg and T.G. Pottinger. 2008. Functional genomics of stress responses in fish. Reviews in Fisheries Science. 16(S1): 157–166.

Scherb, H. and K. Voigt. 2011. Adverse genetic effects induced by chemical or physical environmental pollution. Environ. Sci. Pollut. Res. 18: 695–696.

Selye, H. 1973. The evolution of the stress concept. Am. Sci. 61: 692–699.

Selye, H. 1984. The stress of life. McGraw Hill Publ. 515 pp.

Slaninova, A., M. Smutna, H. Modra, and Z. Svobodova. 2009. A review: oxidative stress in fish induced by pesticides. Neuro Endocrinol Lett. 30 (1): 2–12.

SØrensen, J.G., T.N. Kristensen and V. Loeschcke. 2003. The evolutionary and ecological role of heat shock proteins. Ecology Lett. 6: 1025–1037.

Valavanidis, A., T. Vlahogianni, M. Dassenakis, and M. Scoullos. 2006. Molecular biomarkers of oxidative stress in aquatic organisms in relation to toxic environmental pollutants. Ecotoxicol Environ Saf. 64(2): 178–89.

Vladimirov, Ju. A. and E.V. Proskurina. 2007. Lectures of Medical Biophysics. Moscow University Publ. «Academkniga». 432 pp. (*in Russian*).

Winston, G. 1990. Physicochemical basis for free radical formation in cells: production and defenses. In: Stress responses in plants: adaptation and acclimation mechanisms. R.C. Alscher and Cumming J.R. (eds.). Wiley, New York. 57–86.

Winston, G.W. 1991. Oxidants and antioxidants in aquatic organisms. Comp. Biochem. Physiol. 100 C (1-2): 173–176.

Winston, G.W. and R.T. Di Giulio. 1991. Prooxidant and antioxidant mechanisms in aquatic organisms. Aquatic Toxicol. 19: 137–161.

CHAPTER 2

Biomarkers for Physiological Stress in Fish

Classification, Characterization and Specificity

Studies of the phylogenetic peculiarities of defense systems' evolution in fish are very important for understanding the adaptive mechanisms to damage factors of the environment especially anthropogenic pollution which play a key role in aquatic ecosystems transformation and degradation in recent years. In the organism, xenobiotics bind to specific cellular receptors localized on the membranes and in cytoplasm or in subcellular organelles and cause the negative events in cells and tissues, damage organ and organism itself. Biochemical markers respond to the toxic activity of pollutants, detect the type of toxicity and in some cases the level and vector of response correlate with the level of pollution (Adams, 2005; Amado et al., 2006a,b; Stroka and Drastichova, 2004). According to the WHO (1993), biomarkers can be subdivided into three groups (van der Oost et al., 2003):

"Biomarkers of exposure: covers the detection and measurement of an exogenous substance or its metabolite or product of an interaction between a xenobiotic agent and some target molecule or cell that is measured in a compartment within an organism;

Biomarkers of effect: includes measurable biochemical, physiological or other alterations within tissues or body fluids of an organism that can be recognized as associated with an established or possible health impairment or disease;

Biomarkers of susceptibility: indicates the inherent or acquired ability of an organism to respond to the challenge of exposure to a specific xenobiotic substance, including genetic factors and changes in receptors which alter the susceptibility of an organism to that exposure."

Biomarkers are broadly defined as a change in a biological response (ranging from molecular, subcellular, cellular, physiological and morphological responses) that can be related to exposure to toxic effects of environmental chemicals (van der Oost, 2003; Hutchinson et al., 2006). Responses of the organism on stress insult can be grouped into three classes of indicators including the parameters of various biological levels (Adams, 2005) (Table 2.1).

Biomarkers are selected for their responses to particular classes of the environmental stressors rather than to specific individual chemicals (Goksoyr et al., 1996; Galloway, 2006; Sarkar et al., 2006) (Fig. 2.1).

Fish are very sensitive to anthropogenic impact and some of them may be tested as biomonitors for the assessment of the ecological status of aquatic environment. Comparative study of fish detoxification, metabolism and resistance to stress leading to the anthropogenic pollution is very important for understanding the different strategies of adaptations of aquatic organisms and for risk assessment.

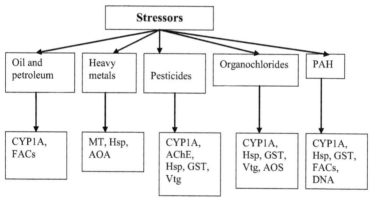

Fig. 2.1. Possible indicative biomarkers on environmental stressors (adopted from Goksoyr et al., 1996; Galloway, 2006; Sarkar et al., 2006). CYP1A—Cytochrome P450 1A, GST—glutathione–S-transferase, FACs—fluorescent aromatic compounds in the bile, MT—metallotioneins, Hsp—heat shock proteins, AChE—acetylcholinesterases, Vtg—vitellogenin, AOS—antioxidant system.

Table 2.1. Parameters characterized by stress responses of the organism (Adams, 2005).

Direct indicators of exposure (biomarkers)	Direct indicators of effects (bioindicators)	Indirect indicators of exposure/effects
Biotransformation system DNA damage	Lipid metabolism Enzyme alterations in organs and tissues	Diet and feeding Fat storage
Antioxidant enzymes Metallotioneins Stress proteins	Immunity Hystopathologies Steroid hormones	Growth Reproduction Behavior

2.1 Biotransformation Systems

Biotransformation is the key process of interaction between xenobiotics and biological systems. There are some pathways of foreign compounds metabolism in organism in which they modify or degrade. Generally, metabolism of xenobiotics includes two phases. The first phase of biotransformation involves the reactions of oxidation, reduction and hydrolysis which may result in the formation of xenobiotics with reduced toxicity or produce active metabolites. In second phase intermediates produced in the first phase and the original chemicals are conjugated with the endogenous compounds (glucuronate, glutathione, etc.) and excreted from the organism (Stegeman and Kloepper-Sams, 1987; Stroka andDrastichova, 2004).

2.1.1 Phase I

The most important oxidative enzymes of the first phase of biotransformation are the family of cytochrome P450 (CYT P450) or "mixed function oxidases" (MFO) which were recognized both in mammals, lower vertebrates and invertebrates (van der Oost et al., 2003; Klemz et al., 2010; Sole et al., 2009). High level of cytochromes was detected in the liver which was estimated as 1–2% of hepatocyte mass. They are present in the intestine, kidney, lung, brain, skin, prostate gland, placenta, etc. The family of cytochrome P450 includes approximately 1000 isoforms. They are membrane-bound proteins, located in the endoplasmic reticulum and are joined to the common basic structure heme. Some of them participate in metabolism of foreign compounds (CYP1-3), metabolism of long-chain fatty acids (CYP4), thromboxane biosynthesis (CYP5), steroid metabolism (CYP7), and steroid biosynthesis (CYP11, 17, 19, 21), prostacycline biosynthesis (CYP8), metabolism of vitamin D (CYP24), retinoids (CYP26), and biosynthesis of bile acids (CYP27). Some isoforms were found in different species of following insects (CYP6 and 9), mollusks (CYP10), fungi (CYP51-70), plants (CYP71-100) and bacteria (CYP101-140). In addition, the isoforms of xenobiotics biotransformation (CYP1-3) can be involved in endogenous metabolism of melatonin and estradiol, testosterone, catecholamines, progesterone and arachidonic acid (Stroka and Drastichova, 2004). Cytochrome P450 aromatase (CYP19) is recognized as the terminal enzyme in the steroidogenic pathway that converts androgens (testosterone) into estrogenes (estradiol). Regulation of this gene regulates the ratio of androgens to estrogenes and the expression of this enzyme is critical for reproduction and sex differentiation (Trant et al., 2001).

Cytochrome P450 enzymes are associated with microsomal fraction that transfers electrons to reduce cytochrome P-450 during its catalytic cycle. These components include NADPH-cytochrome P450 reductase (P450 RED), cytochrome b_5 (CYT b_5) which functions in fatty acid desaturation and NADPH-cytochrome b_5 reductase (Stegeman and Kloepper-Sams, 1987). During the phase I biotransformation via MFO system lipophilic xenobiotics transfer to water soluble compounds which may be excreted via renal system.

Mitochondria also contain monooxygenase CYT P450 complexes. They involve in the metabolism of cholesterol and biosynthesis of steroid hormones, bile acids and vitamin D_3. Besides that CYT P450 family participates in many reactions of hydroxylation, dealkylation, deamination, sulphoxidation, N-oxidation, dehalogenation, etc. (Stroka and Drastichova, 2004).

Phase I biotransformation cycle includes several steps (van der Oost et al., 2003):

Step 1—substrate binds to prostethic heme ferric ion (Fe^{3+}) of the enzyme, the iron is reduced by electron transfer from NADPH cytochrome P450 reductase (P450 RED) and O_2 is bound;

Step 2—addition of a second electron via cytochrome b_5 and the formation of a peroxide, cleavage of the O-O bond, the formation of substrate radical, the hydroxylation of this radical and release of the product.

Great interspecies differences of CYT P450 were documented, but sometimes the similar substrates can be metabolized by different CYT P450 isozymes. CYT P450 isozymes are indicated in tissues of many fish species. The similarity of responses suggests that similar isozymes of CYT P450 are being regulated in teleosts and mammalian species. However, generally microsomal CYT P450 activity in fish was significantly lower than in mammals. They differ as compared to mammals which could be attributed to the lack of structural genes for CYT P450 isoforms related to the mammalian, or to the differences in regulatory processes for such genes. On the other hand various fish and mammalian isozymes are counterparts in these species (Stegeman and Kloepper-Sams, 1987). The overview of the responses of phase I-related enzymes to xenobiotic in both laboratory and field studies of fish was summarized in the publication of van der Oost et al. (2003).

Enzyme induction is initiated by the binding of a specific xenobiotic or protein complex that comprises the AhR and the heat shock protein 90. The AhR complex then binds to aryl hydrocarbon nuclear transferase (ARNT–AhR nuclear translocator) and migrates to the cell nucleus where ARNT binds to a DNA recognition sequence upstream of the CYT P450 genes, also known as the xenobiotic regulatory element (XRE) or dioxine responsive element (DRE). Transcription factors now have ready access to the promoter region of the CYP1A gene. Consequently, mRNA synthesis increases and enzyme level increases also (Stegeman and Hahn, 1994; van der Oost et al., 2003).

Induction of CYT P450 in fish depends upon many biotic and abiotic factors. In several marine fish the levels of CYT P-450 and their catalytic function are 10-fold higher in mature season in male than in female. Estradial suppresses the levels of hepatic CYT P-450 and some catalytic functions in fish and could be the major effector of sex differences in fish as well as in mammals (Stegeman and Kloepper-Sams, 1987).

The isoforms of CYP1, CYP2 and CYP3 are also involved in endogenous metabolism of melatonin and estradiol (CYP1A), testosterone (CYP3A), catecholamines (CYP2D), progesterone (CYP2C, CYP3A) and arachidonic acid (CYP2E). In fish the most sensitive indicator to environmental pollution is CYP1A subfamily (Stegeman and Hahn, 1994). The level and activity of CYP1A in fish is directly related to levels of aromatic and polychlorinated aromatic hydrocarbons (PAH) in the organism and in the environment. CYP1A is responsible for the biotransformation of a myriad of organic compounds (PCBs, dioxins, PAHs, etc.). However, it responds to extremely low concentrations of xenobiotics in water and at the presence of heavy metals (Cu, Zn, Cd, Pb and Ni) CYP1A inhibits resulted in the suppressed expression of the Ah receptor (Stroka and Drastichova, 2004). Hence it is not always possible to have a linear dose-response relationship between the concentration of certain pollutants and CYP1A concentration and/or activity in the natural environment as there is a mixture of both inducers and inhibitors of CYP1A which are likely to act simultaneously (Sarkar et al., 2006).

Generally the total Cytochrome P450 enzymes do not demonstrate the response to pollutants in fish while the specific isozymes are more sensitive to individual pollutants which can induce or inhibit them. Van der Oost and co-authors (2003) summarized the results obtained in 39 laboratory studies and 35 field studies and they observed that significant increase in CYP450 levels was observed in 53% of the laboratory studies and 51% of field studies, while strong increase (> 500% of control) was noted in 3% and 6% of the laboratory and field studies respectively. The researchers concluded that the application of CYP450 response as biomarker for environmental risk assessment (ERA) is limited as compared with the specific forms such as CYP1A, which is the most sensitive to xenobiotics. The CYP1A responses for all fish species from 60 laboratory studies and 48 field studies demonstrated that a significant increase in its levels was observed in 91% of the laboratory studies and 85% of the field studies, while strong increase (> 500% of the control) was observed in 43 and 39% of the laboratory and field studies respectively.

The common method to examine the response of the CYP1A isoenzyme is to determine its catalytic activity, which can be measured by the aryl hydrocarbon hydroxylase (AHH) activity which catalyzes the hydroxylation of benzo[a]pyrene or activity of ethoxyresorufin O-deethylase (EROD), which is more sensitive (Sarkar et al., 2006). Increase in both AHH and EROD activities have been shown in many fish species exposed to organic compounds. A significant increase of AHH activity was observed in 88% of the laboratory studies and 90% of field studies, while strong increase (> 500% of the control) was observed in 43 and 39% of the laboratory and field studies respectively. The EROD activities in fish species from 137 laboratory studies and 127 field studies demonstrated the significant increase in 88% of laboratory studies and 90% of field studies, while strong increase (> 500% of the control) was observed in 69 and 37% of the laboratory and field studies respectively (van der Oost et al., 2003).

The next member of CYT P450 family CYT b_5 measured in fish in 12 laboratory studies and 10 field studies has been shown the increase in 33% of the laboratory studies and 50% of the field studies. No significant increase (> 500% of the control) was observed and the use of this parameter as stress biomarker is questionable. Generally no significant differences of P450 RED activity were shown between fish from polluted and control sites, but the present results are contradictory and the information is very limited (van der Oost et al., 2003).

Hence hepatic CYP1A protein levels and both AHH and EROD activities in fish should preferably be measured in fish for toxicological evaluation of organic compounds because they are very sensitive biomarkers and may be applied in monitoring and ERA processes.

2.1.2 Phase II

Second phase of xenobiotics or the metabolites biotransformation involves the reactions of their conjugation with endogenous ligands, which covalently bind to foreign compounds. The major pathways for electrophilic compounds and metabolites is conjugation with glutathione (GSH), while for nucleophilic compounds conjugation with GA (glucuronic acid) is the major rout. Sulphatation is also one of the pathways of the conjugation reactions, but it is effective at very low substrate concentrations (George, 1994). The mechanism of induction for most forms of phase II enzymes regulated via the AhR, but as compared with phase I system the response of conjugations enzymes is generally less and they may be masked by different exogenous (nutrition, season, temperature, oxygen concentration, salinity) and endogenous (maturity, sex, developmental stage) factors. Heavy metals ions are characterized by an extremely high affinity for SH groups and GSH has been shown to form GS-Me complexes with various metals and GSH is capable of complexing and detoxifying heavy metals cations soon after they enter the cells (Canesi et al., 1999).

Reduced glutathione (GSH) participates in conjugation of electrophilic intermediates via GST activities in phase II and it is an important antioxidant (see section 2.2). The most direct effect of certain xenobiotics is a decrease in the ratio of reduced oxidized glutathione (GSH/GSSG) caused direct radical scavenging or increased peroxidase activity. In mammals GSH synthesis is regulated via feedback inhibition by GSH on rate-limiting synthetic enzyme. In the normal conditions the ratio GSH/GSSG is very high and estimated as 10/1 (Stegeman et al., 1994). In polluted conditions due to exposure to organic compounds EROD induction was enhanced in GSH-supplemented tissues while the reverse was observed in GSH deficient fish. The relations between CYP1A gene expression and catalytic activity and thiol status of fish tissues were observed (van der Oost et al., 2003).

In our studies the interspecies differences of the total glutathione content in fish tissues as well as interspecies variations have been observed in red blood cells (Fig. 2.2) (Rudneva, 2012). Glutathione concentration in fish red blood cells ranged between 22.8 and 60.0 μg%, but the differences did not depend on fish phylogenetic and ecological status. In muscle, liver and gonads the interspecies variations of the total glutathione content were less than in blood.

Previously we described the variations in blood antioxidant system of some Black Sea elasmobranch and teleosts which reflected adaptive strategy of fish species to oxidative stress and their ability to cope with the environment (Rudneva, 1997). High glutathione level in red blood cells in marine elasmobranch has been found and it could compensate for the low

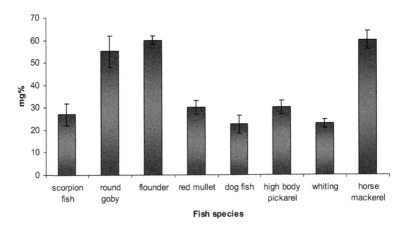

Fig. 2.2. Glutathione concentration in Black Sea fish blood (mg%, mean ± SEM) (Rudneva, 2012).

level of their enzymatic status (Rudneva, 1997; Rocha-e-Silva et al., 2004). Glutathione in erythrocytes protects the hemoglobin against spontaneous oxidation to metahemoglobin and xenobiotics damage. Erythrocytes depleted of glutathione become very sensitive to environmental stress because fish hemoglobin has a higher tendency to oxidation as compared with other vertebrates (Li et al., 2006).

The data of glutathione levels' responses in fish species from 10 laboratory studies and 17 field studies was summarized in the review of van der Oost et al. (2003). A significant increase in GSH levels was observed in 60% of the laboratory studies and 35% of field studies. A significant increase in GSSG levels was observed in two of the laboratory studies (67%) and two of the field studies (50%). Strong increase (> 500% of the control) was detected only in the liver of English sole caught in highly polluted area. The authors have concluded that hepatic GSH and GSSG levels in fish cannot yet be considered as valid stress biomarkers for ERA purposes, but could be used as a potential biomarkers for oxidative stress (van der Oost et al., 2003).

Glutathione-S-transferases (GSTs) E.C.2.5.1.18. are a family of intracellular enzymes with the main function in detoxification processes by catalyzing the conjugation of tripeptide glutathione (GSH) with some endogenous toxic metabolites and many environmental chemicals (Habig et al., 1974; Egaas et al., 1995). The enzymes take part in transport of endogenous hydrophilic compounds, including steroids, heme, pigments, bile acids and their intermediates. In addition, they also play an important role in the detoxification of lipid peroxides and demonstrate the functions such as glutathione peroxidase activity towards reactive oxygen species in the cells in the case of oxidative stress. GST plays an important role in defense against oxidative damage and prooxidative products of DNA and lipids. On the basis of substrate specificity, immunological cross-reactivity and protein sequence data, the family of soluble GSTs has been grouped into four classes: a, m, p and q (George, 1994; Leaver et al., 1993).

Induction of GST activity in aquatic organisms has also been found in high polluted marine environments after the wreckage and oil spills of the tankers (Martinez-Gomez et al., 2009). The increase of GST activity was noted in some fish species and invertebrates collected in environments impacted by complex discharge of contaminants and accidents (Hamed et al., 2003). Hence, the enzymes could be used as a biomarker of water and sediments contamination. GST response among species was clearly quite different. It depends on the peculiarities of their phylogenetic position, biology, physiology and ecological status of the locations (Martinez-Alvarez et al., 2005). The interspecies variations of enzyme activity were shown between fish species (Hamed et al., 2003, 2004; Martinez-Gomez et al., 2009). GST activity in fish liver has been suggested as biomarker of organic pollution of water environments (Filho et al., 2001; Ahmad et al., 2004; Farombi et al., 2007; Hansson et al., 2006; Sun et al., 2006).

GST activity between species was clearly quite different and reflected the peculiarities of their phylogenic position (Filho, 2007; Filho et al., 2001) and diverse ecophysiological characteristics (Sole et al., 2009). The interspecies variations of enzyme activity were shown between fish species (Hamed et al., 2003, 2004). Liver of vertebrates exhibits a high metabolism and oxygen consumption and it is the main organ of xenobiotic detoxification. It is particularly rich source of GST. In rats GST represents some 10% of soluble hepatic protein and in humans about 3% of it (Egaas et al., 1995). In our study GST activity in blood of elasmobranch and teleost fish classes was lower than in liver. Taking into account the peculiarities of metabolic pathways in elasmobranch as compared with teleosts we could propose the presence of specific GST activity in them which agrees with the results of several investigators (Filho, 2007). But we could not postulate that hepatic GST activity in elasmobranch significantly differed related to teleosts and perhaps, the ecological peculiarities of fish species play more important role in GST induction (Rudneva et al., 2010a,b).

Significant interspecies differences were evidenced, but they varied independently in blood and in liver (Rudneva et al., 2010a,b) (Fig. 2.3). The differences in the GST levels in examined fish tissues (blood and

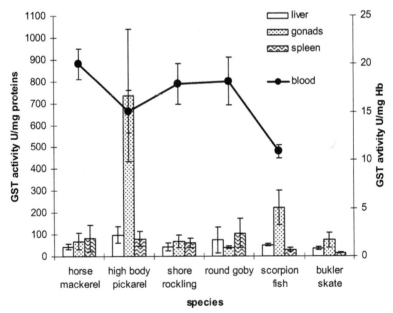

Fig. 2.3. GST activity in tissues (nmol conjugates mg protein^{-1}min^{-1}) and blood (nmol conjugates mg hemoglobin^{-1}min^{-1}) of some Black Sea teleosts and elasmobranch (adopted from Rudneva et al., 2010a,b).

liver) probably indicated that different isozymes are involved, as had been reported for the mammals GST (Eggas et al.,1995), birds (Dai et al., 1996) and fish species (Goksoyr et al., 1996). Study of different classes of isozymes and the sorts of substrates for GSTs in fish is important for understanding the conjugation mechanisms of xenobiotics of ecological interest and comparing them with the other taxonomic groups because hepatic GST activity has been widely used as biomarker of pollution response. Its induction associates with the metabolic function of the liver and its ability to eliminate the oxidative by-products. Differences in hepatic GST isozymes in marine fish from areas with different levels of contamination were reported (Goksoyr et al., 1996; van der Oost et al., 2003).

Our studies of GST activities in blood, liver, gonads and spleen in six Black Sea teleost fish species have shown no significant interspecies differences in examined tissues. A significant interaction between GST levels in blood and spleen was estimated which could be related to the production of the blood cells in the spleen. On the other hand high variability of the values could reflect the individual responses of fish species to environmental stressors (Martinez-Gomez et al., 2009). We did not observe the correlation between fish ecological status and GST activity in liver, gonads, spleen as well as in blood cells. However, we could propose that the differences in the GST activities in examined fish tissues probably indicate that different isozymes are involved, as had been reported for the mammals GST (Egaas et al., 1995) and fish (Nimmo, 1987).

In addition, the interspecies variations of GST levels ion tissues of examined Black Sea fish species may be the result of the different fish sensitivity to organic pollution. Several investigators demonstrated the highest GST activities in hepatopancreas in aquatic invertebrates and in fish liver. These organs play an important role in the detoxification processes for xenobiotics and endogenously generated metabolites that could not be metabolized by the other organs. In this case the authors observed that the suitable biomarker is hepatic GST. At the same time in birds the highest GST activity was identified in the kidney followed closely by that in the liver (Dai et al., 1996). The tissue specific damage corresponded to the differences in the GST activity potentials of the tissues for their adaptation to environmental stress (Ahmad et al., 2004). The interspecies variations of GST activity in fish tissues may reflect the specific adaptations to oxidative stress and defense mechanisms against oxidative damage.

GST activity has been detected in the liver of many elasmobranch and teleosts fish species. In fish as in mammals and birds GST account for an appreciable proportion of soluble hepatic proteins. However, the information of its rations differed and sometimes it's a problem to compare the activities between organisms because of differences in assay conditions and substrates. Additionally, the enzyme does not conjugate with some substrates and in such cases it will not be detected. The most convenient substrate is 1-chloro-2,4-dinitrobenzene (CDNB) because it usually conjugates in high rate (Nimmo, 1987). But the catalytic GST activity towards CDNB is also considered to represent an integration of the activities of most GST classes, and the induction of specific GST isozymes may be masked (Goksoyr et al., 1996; Stephensen et al., 2000).

We could propose that GST plays an additional role in the detoxification process of scanty xenobiotcs in erythrocytes while the main mechanism is provided by passive gill excretion of pollutants. Such a mechanism was described by Filho et al. (1993) for the explanation of the absence or relatively low concentration of glutathione peroxidase in some fish species which could get rid of H_2O_2 by using passive gill excretion also.

No clear relationships between fish swimming capacity and blood GST activity could be established for the totality of species investigated in our studies. Our results demonstrated the higher GST activity in the liver of sluggish and slow swimming fish species (with the exception of scorpion fish) in relation to active forms. Taking into account that most of the sluggish fish species belong to benthic and suprabenthic groups we could propose that they live in more contaminated environment because many pollutants accumulate in bottom sediments and low water layers. Thus, hard pressing of chemicals induces hepatic GST activity especially in benthic and suprabenthic forms.

Fish trophic level, feeding behavior and nutrition factors also may affect biomarkers including GST (Sole et al., 2009; Martinez-Alvarez et al., 2005). In our studies of Black Sea fish species belonging to different feeding groups no clear relationships between feeding classes and GST activities in blood and liver were shown. However, carnivorous fish species demonstrated approximately similar enzymatic activity in the liver. It could be explained that benthic invertebrates (mollusks, crustacean and worms) and fish which are the preferable prey for such a group might accumulate xenobiotics from the bottom sediments and transfer them via trophic nets to fish tissues. Hepatic GST activity in pelagic plankton feeder horse mackerel and omnivorous suprabenthic/pelagic pickerel and golden grey mullet with the exception of peacock wrasse was approximately similar and it was lower than in carnivorous, which was explained by the similarity of their high metabolic rate and diet consumption. We found high variability of hepatic GST activity among predators, but in blood the values varied less (Rudneva et al., 2010a,b). The results of our study showed that the highest GST activity was noted in red blood cells in fish caught in the most polluted site and it was 1.5-2-fold higher than in fish from both reference and lower polluted bays (Fig. 2.4) (Rudneva et al., 2012).

In the review of van der Oost et al. (2003) the GST responses for all fish obtained from 43 laboratory studies and 39 field studies were

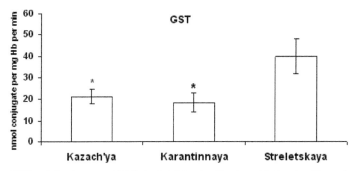

Fig. 2.4. GST activity (mean ± SEM) in fish blood cells. Asterisk *—significant (p < 0.05) differencein the values of the Streletskaya Bay.

summarized. The significant increase in GST activity was observed in 33% of the laboratory studies and in 33% of the field studies and no strong increase (> 500% of the control) was reported. The authors have suggested that the total hepatic GST activity in fish was not used as biomarker for ERA, but the specific isozymes are more sensitive and selective to organic pollutants such as PCD and PAHs (van der Oost et al., 2003). The authors also suggest that the interspecies variations of GST activity may reflect the specific adaptations to the oxidative stress and protective mechanisms against oxidative damage. Thus, the analysis of GST activity in different fish species taking into account their phylogenic position, specific features of ecology and physiology is important for the evaluation of fish abilities to protect against pollutants and keep their life and biodiversity in the impact environments (Rudneva et al., 2010a,b).

The synthesis of glucuronides by microsomal UDP-glucuronyl transferases (UDPGTs) is the additional pathway for organic endogenous (steroid hormones) and exogenous compounds biotransformation (Morcillo et al., 2004). This could affect excretion and consequently, endogenous levels of steroid hormones, which may produce a cascade of undesirable effects for the organism. This family involves many isozymes which are generally named after their acceptor substrates (George, 1994; van der Oost et al., 2003). The diversity of UDPGT was also detected in fish tissues and its activity depends on sex, season, pH and temperature (Stegeman and Lech, 1991; Stegeman and Hahn, 1994). The analysis of UDPGT responses for all fish species from 27 laboratory studies and 26 field studies has indicated that a significant increase in enzyme activity was observed in 52% of the laboratory studies and in 42% of the field studies, while strong increase (> 500% of the control) was reported only in two laboratory studies. The researchers have concluded that UDPGT activity is the most responsive to pollutant exposure among phase II parameters and may be used as a good biomarker in ERA procedure (van der Oost et al., 2003).

Hence, the results obtained by several researchers and our findings show that the induction of biotransformation system and its characteristics in fish depend on species biology. The complexity of specific phylogenic, physiological and ecological features of fish species may influence the phase I and phase II members and it is important for development of monitoring programs. As we see biomarkers of phase II in benthic forms are more convenient for monitoring studies, but benthic fish species are different from each other, suprabenthic forms are more homogeneous when compared with suprabenthic/pelagic and benthic fish species. However, among suprabenthic species we found the forms with different swimming capacity and type of feeding. Generally the studies of biotransformation system in fish show that the complex of biotic and abiotic factors including anthropogenic impact may be attributed to use these parameters as stress biomarkers.

2.2 Antioxidant System and Lipid Peroxidation

Metabolism of the living organisms is associated with oxygen consumption involving it into the major pathways. However, O_2 has two unpaired electrons and the univalent reduction of the molecular oxygen produces reactive oxygen species (ROS) such as superoxide radical (O_2^-), singlet oxygen ($^1O_2$), hydrogen peroxide (H_2O_2), hydroxyl radical ($HO^{\cdot}$) and finally water H_2O. Nitric oxide ($NO^{\cdot}$) may also be formed in the metabolic processes. ROS may be harmful for biological systems because they cause oxidative stress, the generation and accumulation of ROS in tissues which damage cell membranes, lipids, proteins, and DNA. Lipids and particular polyunsaturated fatty acids are the main substrates of the oxidation and lipid peroxidation (LPO) is a very important consequence of oxidative stress. On the other hand ROS can also act in signal transduction and participate in the formation of intermediates (prostaglandins, leikotriens, etc.), and play an important role for the expression of several transcription factors, heat shock-inducing factor, nuclear factor, the cell-gene p53, nitrogen-activated protein kinase, etc. Oxidative stress also plays a role in apoptosis in two pathways, the death-receptor and the mitochondrial (Winston, 1991; Livingstone, 2001; Burlakova, 2005; Lesser, 2006; Vladimirov and Proskurina, 2007).

Antioxidant system plays a key role in inactivation of reactive oxygen species (ROS) and thereby controls oxidative stress as well as redox signaling. Organisms have antioxidant defense to counteract the toxic effects of ROS. Antioxidant defense of organisms depends on many abiotic (temperature, season, salinity, oxygen concentration) and biotic (feeding conditions, diet quality and quantity) and other environmental factors (Martinez-Alvarez et al., 2005). Antioxidant status of the organism is detected by interactions between prooxidant and antioxidant processes and the presence of the antioxidants (Fig. 2.5). Antioxidant system plays an important role in all living organisms and it includes low molecular weight scavengers and special adopted enzymes.

Low molecular weight antioxidants are non-enzymatic compounds such as vitamins A, E, K and C, carotenoids, SH-containing amino acids and peptides (glutathione) and small-molecule antioxidants are uric acid, urea, etc (Winston and Di Giulio, 1991). Ascorbic acid functions as reductant source for many ROS, it scavenges both H_2O_2 and O_2^-, HO^*, and lipid hydroperoxides without enzyme catalysts. Additionally it plays an important role in recycling α–tocopherol to its reduced form (Lesser, 2006). Glutathione (GSH) is a tripeptide (GLU-CYS-GLY) which forms a thiol radical that interacts with a second oxidized glutathione and forms a disulphide bond (GSSG). The ratio of GSH/GSSG is used as a biomarker of oxidative stress in the organism (Livingstone, 2001).

GSH oxidizes H_2O_2 and organic hydroperoxides and forms GSSG. This process is spontaneous or catalyzes by the enzyme glutathione peroxidase. GSSG reduces again to GSH via reaction

$$GR\text{-}GSSG + NADPH^+ + H^+ \rightarrow 2GSH + NADP^+.$$

Serum SH-groups of the amino acids also participate in antioxidant defense of fish. Their level ranges in serum in different species and depends on their taxonomic position. It was obsereved that the content of the total SH-groups was significantly higher in the serum of Black Sea elasmobranch *Squalus acanthias* (p < 0.01) as compared to the teleosts and the level of non-protein SH- groups was greater than the protein ones (Rudneva, 1997, 1998). GSH concentration in fish serum also varies which has been described above (see section 2.1).

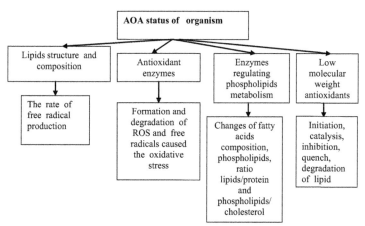

Fig. 2.5. The links between prooxidant and antioxidant processes in the organism (Rudneva, 2012).

Tocopherols, especially α–tocopherol (vitamin E) are lipid-soluble antioxidants which are located within the bilayers of cell membranes and protect them against ROS (Burlakova, 2005). It is multifunctional antioxidant which is the result of its ability to quench both $^1O_2$ and peroxides. Carotenoids are also lipid-soluble compounds that protect against $^1O_2$ because they have highly conjugated double bonds and they quench ROS and can prevent lipid peroxidation in marine animals (Winston, 1990; Hussein et al., 2006). Additionally, carotenoids are the precursors of vitamin A which can be a potent antioxidant (Lali and Lewis-McCrea, 2007). Vitamin K protects the organism against ROS also (Hardy, 2001).

Uric acid can quench both $^1O_2$ and HO. It is found in high concentrations in marine invertebrates and in blood serum of elasmobranch species and it can be a potent antioxidant (Lesser, 2006). There are some other low molecular weight antioxidants in aquatic organisms but in some cases their nature is unknown.

Antioxidant enzymes play an important role in inactivation of reactive oxygen species and thereby control oxidative stress as well as redox signaling. Both processes change across the life span of the organism and thus modulate its sensitivity and resistance against free radical damage (Vifia et al., 2003; Hu et al., 2007). Enzymatic antioxidant system includes some enzymes which catalyze the reactions of ROS degradation. Superoxide dismutase (SOD) (EC 1.15.1.1.) protects against oxidative damage by catalyzing the reaction of dismutation of the superoxide anion to H_2O_2:

$$2O_2^- + 2H^+ \rightarrow H_2O_2 + O_2$$

Catalase (CAT) (EC 1.11.1.6.) catalyzes the reduction of hydrogen peroxide to water

$$2H_2O_2 \rightarrow 2H_2O + O_2$$

Selenium-dependent glutathione peroxidase (SeGPX) (EC 1.11.1.9) catalyzes the reaction of hydroperoxides reduction:

$$H_2O_2 + 2\ GSH \rightarrow 2\ H_2O + GSSG$$

Total peroxidase (PER) (the sum of SeGPX and selenium-independent GPX activities) reduces both hydrogen peroxide and organic hydroperoxides:

$$ROOH + 2GSH \rightarrow ROH + H_2O + GSSG$$

Glutathione reductase (GR) (EC 1.6.4.2.) catalyzes the reaction

$$NADPH + H^+ + GSSG \rightarrow NADP^+ + 2GSH$$

and maintains a ratio GSH/GSSG under oxidative stress.

Quinon oxidoreductase NAD(P)H-dependent DT-diaphorase (EC 1.6.99.2) catalyzes the two electron reduction of redox cycling quinones and related compounds to hydroquinones so preventing their univalent reduction to quinine anion radicals leading ROS production via autooxidation (Peters and Livingstone, 1996):

$$NAD(P)H + Q + H^+ \rightarrow QH_2 + NAD(P)^+$$

where Q and QH_2 are respectively quinone and hydroquinone.

Antioxidant status of fish depends on many abiotic factors such as temperature, season, diet, salinity, oxygen concentration and many others, as well as, on their taxonomic position and life cycle. Previously we described the variations in blood antioxidant system of some Black Sea elasmobranch and teleosts which reflected adaptive strategy of fish species to oxidative stress and their ability to cope with the environment (Rudneva, 2012). Several investigators also documented the lack or relatively low contents of CAT and GR activity in blood of some fish species which was led due to the presence of high levels of H_2O_2 in the blood (Filho and Boveris, 1993; Filho et al., 1993, 2005).

The researchers also reported that antioxidant status can be correlated with phylogenic position and ecological peculiarities and these parameters in primitive life are less than in mammals and birds (Dafre and Reischl, 1990; Rudneva, 1997; Rocha-e-Silva et al., 2004; Sole et al., 2009). In our study we displayed the absence of CAT activity in blood red cells of Black Sea elasmobranch *S. acanthias* (Rudneva, 1997, 1998). The levels of antioxidants and antioxidant enzymes activities were ranged in fish tissues. Our findings showed that the concentration of vitamin A in blood serum and glutathione in red cells was differed approximately 2-fold in Black Sea fish species. High contents of low molecular weight antioxidants were detectable in fish muscle (Rudneva, 2012). Hepatic antioxidant enzymes level varied widely and the comparative study of the values of liver and muscle enzymatic activities did not show uniform trends. In contrast the authors documented the highest SOD and CAT levels in fish liver as in high metabolic rate organ (Filho et al., 1993; Zelinski and Portner, 2000).

Thus taking into account the specificity of metabolic pathways in elasmobranch as compared to teleosts we could propose the following compensatory mechanisms in their antioxidant defense (Rudneva, 2012):

- *in blood*: the high concentration of SH-groups especially non-protein;
- *in muscle*: high concentration of glutathione and vitamin K, high level of PER which is compensated by low level of CAT;

- *in liver*: high concentration of vitamin K and high value of total lipid AOA which is associated with the presence of fat-soluble antioxidants of unknown nature;
- *in gonads*: high levels of glutathione, vitamin K and E and total lipid AOA in male; development of the embryo in maternal organism which protects it against oxidative stress of the environment.

Despite this we could suggest that ecological peculiarities of fish species play an important role in antioxidant defense than phylogenetic position. Our study supports the great interspecies differences in antioxidant activities in tissues between various fish species, which agrees with the results of other researchers, who reported about the high variability of antioxidants levels in fish (Filho et al., 1993; Sole et al., 2009; Winston, 1991). Our findings demonstrated that the interspecies differences in blood, muscle and liver in Black Sea fish species were more significantly than in gonads which showed more homogenous response (Rudneva, 2012).

The interspecies variations of AOA defense may reflect the specific adaptations to the oxidative stress and protective mechanisms against ROS damage. Thus the activity of antioxidant enzymes in fish blood correlated with their swimming capacity. In our previous study we found that SOD and CAT activities in fast swimming pelagic horse mackerel *Trachurus meditrraneus* and pickerel *Spicara smaris* were significantly higher than in the blood of slow swimming gobies, scorpion fish and flounder (Rudneva, 1997). Other researchers also reported that CAT and SOD contents in blood of more active forms were higher as compared to more sluggish species (Filho et al., 1993). At the same time the changes in the sturgeon blood prooxidant-antioxidant status, as a consequence of adaptation to marine conditions, were not reflected in the liver and other tissues (Martinez-Alvarez et al., 2005).

Liver of the vertebrates exhibits a high metabolism and oxygen consumption and it is the main organ of xenobiotic detoxification. Fish liver displayed the highest levels of the key antioxidant enzymes SOD and CAT (Filho and Boveris, 1993; Filho et al., 1993; Rocha-e-Silva et al., 2004). The SOD and CAT levels in liver appear to indicate that the most active species both teleosts and elasmobranches had greater enzyme activity as compared to low mobile forms (Filho et al., 1993; Filho and Boveris, 1993). The higher activity of antioxidant enzymes in liver of active fish correlated with the higher oxygen consumption in fast swimming species and their high metabolic rate (Filho, 2007; Martinez-Alvarez et al., 2005). Animals with high metabolic rate exhibit a high rate of free radical production and cause the induction of antioxidant defense mechanisms (Zelinski and Portner, 2000).

In recent years the anthropogenic press on natural ecosystems especially aquatic locations led high level pollution. Xenobiotics distributed in water and accumulated in sediments and biota cause oxidative stress in aquatic organisms and modify the interactions between the main metabolic processes and provoked negative consequences in organisms. Several studies have shown that oxidative stress due to the environmental factors is involved in age-related processes and modulate animals growth, development and metabolism. Oxidative stress leading to immune system dysfunction seems to have an important role in fish senescence (van der Oost et al., 2003; Adams, 2005; De La Fuente et al., 2005).

Measuring of oxidative stress and antioxidant defense characteristics is a good tool in environmental monitoring. Fish antioxidant defense is being used as a biomarker for water pollution and ecological risk evaluation. Pollutants accumulated in organism modify the interactions between the main metabolic processes and influence negatively on the lifespan. Fish liver is the suitable organ for biomarkers determination in most of the monitoring studies (Goksoyr et al., 1996). However, blood is affected by environmental stressors for the first time and serum proteins transport some xenobiotics to the liver and other organs where they accumulate and degrade. In addition, many clinical blood biomarkers were adapted to ecologically relevant purposes in fish and marine invertebrates (Galloway, 2006).

Polluted response differed in various taxonomic groups of animals. Gwozdzinski et al. (1992) have shown that the effect of Cu and Hg on antioxidant enzyme activities in fish red blood cells was significantly higher as compared with the human erythrocytes. The authors have described the modifications of blood antioxidant system of fish affected on chemicals (heavy metals and organic intoxication) and physical stressors (salinity, temperature and oxygen concentration variations) (Roche and Boge, 1996; Martinez-Alvarez et al., 2005; Filho et al., 2005; Ozmen et al., 2004; 2007). Oxidative stress in erythrocytes of Nile tilapia from polluted site was also documented (Bainy et al., 1996). In our previous studies we have observed the effects of anthropogenic pollution on antioxidant status of some Black Sea fish teleosts (Oven et al., 2000, 2001; Rudneva, 2011; Rudneva and Petzold-Bradley, 2001; Rudneva and Kuzminova, 2011; Rudneva et al., 2005, 2008a,b, 2010c).

In Sevastopol region we have several bays characterizing different levels of anthropogenic impact (Rudneva, 2011). We studied the polluted effects of domestic sewage on liver somatic index (LSI) and blood antioxidant status of biomonitor species of scorpion fish *Scorpaena porcus* and the relations between the level of pollution, LSI values and antioxidant status of fish were observed. In Black Sea a number of fish species have the potential of fulfilling "biomonitor" role. We selected scorpion fish *Scorpaena porcus* (Fig. 2.6) because it is among the most common benthic fish species in Black Sea

Fig. 2.6. Black Sea Scorpion fish (*Scorpaena porcus*).
Color image of this figure appears in the color plate section at the end of the book.

coastal waters and it does not migrate. Its populations in tested bays are
stationary and its living history and biology is well known (Rudneva and
Zherko, 2000; Rudneva et al., 2005; Rudneva et al., 2011).

The factor of the total blood antioxidant enzyme activities estimated
as sum of the values of individual enzymatic activity has shown its
dependence on the pollution level in the bay (Table 2.2). Thus the obtained
results demonstrated the correlation between the pollution level and fish
biomarker response in fish. Our findings have shown that the natural
variability of the coastal waters impacted by various anthropogenic
activities including mixtures of municipal, agricultural and industrial
contaminants tend to accumulate them in fish tissues and may lead to the
changes in morphological and biochemical characteristics. We indicated
the increase of LSI in fish from the most polluted bay (Sevastopolskaya).
The similar trend of increase of the total blood antioxidant activities was
shown in fish from the high polluted bays. The induction of antioxidant
enzymatic activities was associated with the defense mechanisms against
oxidative stress induced in the mixtures of pollutants which flows into the
bays. The highest factor was indicated in the fish from the most polluted

Table 2.2. LSI and total blood antioxidant activities in *S corpaena porcus* inhabiting different
Sevastopol bays.

Bays	Sewage level, $m^3\ day^{-1}$	LSI	Total blood antioxidant enzyme activities
Sevastopolskaya	15,000	4.5	67
Balaklava	10,000	3.3	60
Streletskaya	350	3.2	57
Karantinnaya	50	2.1	51
Martynova	-	1.8	40

LSI—liversomatic index, calculated as liver weight x 100/somatic weight, total blood
antioxidant enzyme activities estimated as sum of the values of CAT + SOD + PER + GR.

Sevastopolskaya Bay. SOD activity in fish blood from the most polluted areas in Sevastopol coastal waters (Sevastopolskaya and Streletskaya bays) was significantly higher when compared to fish values from reference sites (Fig. 2.7).

In our study of Black Sea fish species belonging to different ecological groups we observed that the polluted response of AOA system was much more clear in benthic forms as compared with suprabenthic, suprabenthic/ pelagic and pelagic forms (Rudneva, 2011). We propose that benthic forms live in more contaminant environment because many pollutants are accumulated in bottom sediments and in low water layers. Other reasons of this fact are associated with fish trophic level, feeding behavior and nutrition factors which also may affect antioxidant enzymes (Hamed et al., 2003; Martinez-Alvarez et al., 2005; Sole et al., 2009). Benthic invertebrates (mollusks, crustacean and worms) and fish are the preferable prey for this ecological group and they accumulate xenobiotics from the bottom sediments and transfer them via trophic nets to fish. Thus, hard impact of stressors induces high level of enzyme antioxidant activities.

We also reported different types of age-related responses of antioxidant enzyme system in fish belonging to different ecological groups which are not uniform and depend on fish ecological and physiological status (Rudneva et al., 2010c). This could create the problem of the application of these parameters as biomarkers in pollution monitoring of the marine

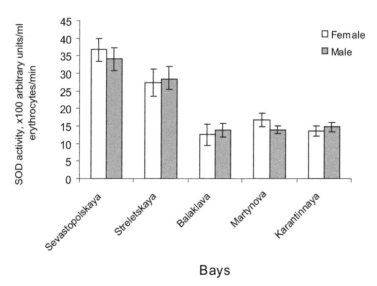

Fig. 2.7. SOD activities in blood (ml erythrocytes[-1]. min[-1], mean ± SEM) in *S. porcus* from Sevastopol' bays with different level of pollution (Sevastopolskaya and Streletskay bays are highly polluted as compared with Bakalava Bay, Martynova Bay and Karantinnaya Bay).

environment. Therefore future investigations should be focused on the search and selection of the reference group of fish which is age homogenous because age-related relations of antioxidant system of fish has already been documented (Otto and Moon, 1996; Weigand et al., 2000).

Studies of antioxidant system in fish is very important for understanding the response mechanisms against pollution in fish of various taxonomic and ecological groups as well as for aquaculture purposes and for receiving the information of repair mechanisms and adaptation to unfavorable impact. Analysis of antioxidant levels in different fish species taking into account their phylogenic position, specific features of ecology and physiology is important for understanding the key ways of evolution of aerobic life, evaluation of fish abilities to protect against pollutants and keep their life and biodiversity in the contaminated environments. Various characteristics of fish biology are reflected on antioxidants level in tissues. Complex of specific phylogenic, physiological and ecological features of fish species may modify antioxidant status and it is important to understand for development of monitoring programs. Antioxidant defense biomarkers of benthic forms are more convenient for monitoring studies, but benthic fish species differed from each other, suprabenthic forms were more homogeneous as compared to suprabenthic/pelagic and benthic fish species. However, among suprabenthic species we found the forms with different swimming capacity and type of feeding. Complex of biotic and abiotic factors including anthropogenic impact may be attributed to antioxidant defense in fish and antioxidant enzymes especially SOD are good tools for the evaluation of pollution effects on fish and the ecological status of their habitats.

2.3 Stress Proteins

The heat shock genes are present in all organisms from bacteria to mammals. High level of conservation and their universal presence reveal their evolutionary and ecologically importance in protection of the cells against stressors. Cellular stress response is regulated with the transcriptional, translational and post-translational levels. The main regulator of the heat shock response genes is the heat shock transcriptional factor 1 (HSF1) and the major activator of it is heat shock. Misfolded proteins titrate out HSF1 form the inhibitory chaperone complex, HSF trimerizes, becomes phosphorilated and is translocated to nucleus where it binds to the heat shock element of Hsps genes. The induced proteins are known as stress or heat shock proteins, which are evolutionary conserved and are present in all organisms and in all cellular compartments (Soti and Csermely, 2007). Molecular chaperones are the large group of proteins that are involved in "house-keeping" functions in the cell. "Chaperone" is adopted from one of their functions, namely to keep other proteins getting involved in "appropriate" aggregations. Several of the major stress proteins are constitutively expressed at low concentrations and function as "molecular chaperones" in the cell to maintain its homeostasis at normal conditions. Their function is related to the binding of Hsps to other proteins or polypeptides and help them to cross cell membranes, to assist in the polypeptide assembling and to prevent anomalous conformations of the proteins, to bind to a variety of intracellular receptors in the absence of ligands. They play an important role in the modulation of immune system also. Hsps are involved in the protection and repair of damaged protein targets and DNA repair (del Razo et al., 2001). Their level increases in different stressful conditions, e.g., in arctic fish Hsps are identified at 5°C and in termophilic bacteria at around 100°C. Hsps play an important role in resistance of the organisms to different kinds of stressors both chemical, physical and biological nature (Serensen et al., 2003).

Ranging in size from approximately 15 kDa to 110 kDa in molecular weight, some of these proteins are constitutive (are found in the cell under normal conditions), while others have been found to be induced in response to a variety of cellular stressors. The elevated expression of stress proteins is the universal response to adverse conditions which reflects the adaptation of the organism to these conditions. They are induced by natural and anthropogenic stressors, including high or low temperature, UV-irradiation, heavy metals, pesticides, oxygen deficiency, anoxia, salinity stress, high density, bacterial and viral infections, parasites invasion, physical activity, oxidative stress (Heikkila et al., 1982; van der Oost et al., 2003), etc. Genetic stressors such as senescence, inbreeding and deleterious mutations also induce Hsps synthesis (Serensen et al., 2003). The induction of stress proteins synthesis is highly specific and depends on target tissues, damage stressors, and other factors. Target-tissue-specific alterations in protein synthesis may represent "biochemical fingerprints" of exposure or "toxicological signature" of toxicity (Sanders, 1993; Sanders et al., 1994; Di Giulio et al., 1989; Hart, 1996; del Razo et al., 2001). They protect cells against damage and repair their structures and comprise the members of different families which are classified according to the homology of their sequences and their relative masses as follows: heat shock proteins (Hsps), glucose-regulated protein (GRP), metallotioneins (MT), ubiquitin (UB), heme oxygenase (HO) (Table 2.3).

Some of the major chaperones (Hsp70, Hsp90, low molecular weight Hsp) are present in normal conditions in the cell reaching 1–5% of total cellular protein for the protein conformational homeostasis. Hsps display various functions in cell, such as proper folding of nascent polypeptide chains, facilitating protein translocation across various cellular compartments, modulating protein activity via stabilization and/ or maturation to functionally-competent conformation, masking mild mutation at the conformational level, promoting multiprotein complex

Table 2.3. Characteristics of stress proteins families (del Razo et al., 2001; van der Oost et al., 2003; Iwama et al., 2004; Soti and Csermely, 2007; Padmini, 2012).

SP families	Localization in cell	Molecular mass, kDa	Factors of induction	Function in cells
Metallotioneins (MT)	Cytoplasm, lysosomes, nucleus	6–7	Transitional metals, dexamethasone, lypopolysaharides	Zinc homeostasis, binding to heavy metals, capture ROS
Ubiquitin (UB)	Cytoplasm, cell surface membrane, nucleus	7–8	Environmental stressors	Cofactor for cytosolic degradation of abnormal proteins to maintain chromosome structure
Heme Oxygenase (HO)	Cytoplasm, mitochondria, endoplasmic reticulum (ER)	32	Heme, metalloporphyrins, transition metals	Physiological breakdown of heme
Glucose-related proteins (Grps)	ER	74–78 94	Glucose starvation And stress responses, disruption of glycoprotein trafficking Pathological processes	Play a role in specialized cell structures, involve in the development of the thermotolerance and sensitivity to chemicals, protect proteins related to autoimmune diseases
Heat shock proteins (Hsps) Hsp27 Hsp 60kDA	Cytoplasm Cytoplasm, mitochondria	16–27 60–65		Prevent the aggregation of misfolded proteins Molecular chaperone, mediation the repair and degradation of altered or denaturated proteins
Hsp 70kDA	Cytoplasm, mitochondria, ER	68–73		Support various components of the cytoskeleton, enzymes and steroid hormone receptors. Coordinate and regulate a variety of diverse signaling proteins and receptors
Hsp 90kDA	Cytoplasm, ER	85–90		Play a role in thermal tolerance caused by strong inhibition of heat-induced protein aggregation
Hsp 110kDA	Cytoplasm, ER	110–170		Not clear

assembly-disassembly, refolding of misfolded proteins, protecting against protein aggregation, targeting ultimately damaged proteins to aggregates, and solubilizing protein aggregates for refolding/degradation (Soti and Csermely, 2007).

In addition, several studies documented a correlation between decrease in cellular energy (ATP) and Hsp activation. ATP depletion is thought to cause accumulation of denatured proteins and aggregation of constitutive Hsp70 that ultimately triggers the heat shock response under non-heat temperature. It has been proposed that during metabolic stress (i.e., decrease in ATP) the constitutive and inducible forms of Hsp70 become stable bound with proteins that are unable to fold or have become denatured. As the available Hsp70 concentrations decrease, the cells respond by increase Hsp expression through a feedback mechanism. At the same time heat shock alone had no significant effect on ATP level but they induced increase gene transcription of the major stress protein Hsp70 in rainbow trout blood cells (Currie et al., 1999).

Synthesis of Hsps is highly increased by exposure of the organism to heat and other physical and chemical stressors. Stress proteins have been recognized as being one of the primary defense mechanisms that are activated by the occurrence of denatured proteins in the cell. Hsp families in comparison to the heat shock-induced proteins have been classified, as follows (Moseley, 1997; del Razo et al., 2001):

- analogous induction, when heat or metals induce the similar proteins;
- subset induction, the proteins induced by metals represent a subgroup of those induced by heat;
- specific induction, when proteins are induced exclusively by metals (metallothioneins, MT), which involves in the detoxification of metals.

Four major stress-protein families of 90, 70 and 60 kDa and low molecular weight Hsp 16–24 kDa are the most prominent and are frequently referred to as Hsp 90, Hsp70, Hsp60 and low molecular weight (LMW) stress-proteins (Bierkens, 2000). LMW are well known as chaperones, they protect the damaged proteins against aggregation in cell, interact with them, transfer them to lysosomes, bind with steroids, defend the t-RNA, and regulate apoptosis (Dhainoaut et al., 1997). The induction of Hsp70 is associated with the presence of damaged proteins, but as other detoxification systems come into play, stress proteins levels return to normal (Vedel and Depledge, 1995). The main characteristics of different Hsp families are present in Table 2.3.

Hsp 27 has the similar structure with αβ-crystallin. Both proteins' level increase in pathological processes, they stabilize denatured proteins, so they can be reactivated by other chaperones or stabilized by some types of actin in response to stress. These proteins participate in the scaffolding of the cytoskeleton and play a role in specialized cell structures as tight junctions between cells, as in the case of the microfilaments of the Sertoli cells, involved in the development of the thermotolerance and sensitivity to chemicals, protect proteins related to autoimmune diseases (del Razo et al., 2001).

The *Hsp 60* is the family of chaperones that participate in synthesis, transportation and degradation of proteins. They prevent the aggregation of misfolded functional proteins. This family comprises several members. Hsp65 is the most important because it is involved in autoimmunity by stimulating antibodies against itself (del Razo et al., 2001; Sanders et al., 1991).

The *Hsp 70* family contains distinct protein members residing in number of subcellular compartments such as cytoplasm, mitochondria and endoplasmatic reticulum (ER). It is one of the major heat shock proteins group that has been studied in various organisms both in laboratory and natural conditions. These proteins act to stabilize or solubilize target proteins and serve a "chaperone" function by helping newly synthesized secretory and organelle proteins to be translocated across the membrane. They prevent incorrect protein folding and translocation binding to the growing peptide chain and maintaining it in a loosely folded state until synthesis is complete. These proteins play a key role in reproductive physiology, gametogenesis, embryogenesis and organogenesis (del Razo et al., 2001).

Hsp 90 is an abundant cytosolic protein in normal cells and is synthesized in stress conditions. The family contains two members: Grp94 (or Grp96) which is mainly resident in the ER and Hsp 90 which is a cytosolic protein. In normal cells Hsp90 coordinates the trafficking and regulates a variety of diverse signaling proteins. It is associated with a number of regulatory proteins including steroid hormone receptors, retinoid receptor, cytoskeleton proteins, calmodulin and βγ subunit of G proteins and forms complexes with several protein kinases (del Razo et al., 2001). The interaction of Hsp 70 and Hsp 90 with stress receptor has been characterized useing cell line models (Iwama et al., 1999).

Hsp110 family contains two subfamily *Hsp110* (in cytoplasm) and Grp170 (in the ER) which are relatives of the Hsp70/Grp78 family. Hsp110 is found in all tissues while its expression is the highest in the brain. Hsp110 is found in conjunction with Hsp70 in the cytoplasm and nucleus and Grp170 is found in conjunction with Hsp70 in the ER. The most distinctive action of the overexpression of Hsp110 is a significant increase in thermal tolerance caused by strong inhibition of heat-induced protein aggregation *in vitro* (del Razo et al., 2001).

Metallotioneins (MT) are low molecular weight proteins which contain approximately 60 amino acids, among them 20 are cysteine and non of them are aromatic occurring mainly in the cytosol and in the nucleus and lysosomes. In some cases MTs contain high cystein content than estimated as 30% (Sarkar et al., 2006). They are non enzymatic peptides. The thiol groups of cystein residues enable MTs to bind particular cations of heavy metals (Sarkar et al., 2006). MT has been associated with zinc homeostasis and metal detoxification by binding to several (7–12) heavy metals. The family comprises four isoforms, MT I and MT II are the major ones which are inducible in many species and tissues. MT III is detected in glial cells and fails to respond to various regulating chemicals such as Zn, Cd, dexamethasone, and others. MT IV is present only in ephithelial cells lining in the gastrointestinal tract (del Razo et al., 2001). Tissues directly involved in metal uptake, storage and excretion have a high capacity to synthesize MTs. In fish these proteins have been identified in high concentrations in gills, liver and some other tissues. MTs bind to excess of essential or toxic metals and protect the organism against toxicity by restricting the availability of these cations at detrimental sites. Some toxic metals (Ag, Cd, Cu, Hg and Zn) have high binding affinity for cystein (Sarkar et al., 2006).

At the other hand several organic pollutants (BaP and PCB) in combination with Cd treatment displayed inhibitory effects on hepatic MT induction in flounder. When BaP was injected prior to Cd, a significant reduction (37%) in MT levels was observed as compared to treatment with Cd alone, and the MT level was the same as in control group. There was also a significant decrease (50%) in MT concentration when PCB-156 was injected prior to Cd, but the levels were still significantly elevated as 23-fold compared to control fish (Goksoyr et al., 1996).

Ubiquitin (Ub) is small molecular weight protein involved in the non-lysosomal degradation of intracellular proteins. The Ub system in mammals comprises of more than 20 proteins from which E2 and E3 are Ub-conjugating and Ub-ligating respectively, which play an important role in cellular regulation processes such as growth control, carcinogenesis and DNA repair (del Razo et al., 2001).

Heme oxygenase (HO) function is associated with the physiological breakdown of heme into equimolar amounts of biliverdin, CO and Fe using O_2 and electron donated by NADH-cytochrome P450 reductase. The extremely increase in heme synthesis results to cell death and its reduction is linked to a decrease in CYP450 inhibiting its microsomal and mitochondrial functions (del Razo et al., 2001). Products of HO reaction have an important effect: HO is a potent vasodilator, which probably plays a key role in the modulation of the vascular tone, especially in the liver under physiological conditions and in many organs under "stressful" conditions. Biliverdin and bilirubin protect cells against oxidative damage as free radical scavengers, in contrast Fe increases ROS level. There are three isoforms of HO (HO-1, HO-2 and HO-3), where HO-1 is the major form while HO-2 is present in brain and testes and HO-3 is probably involved in heme binding.

The next major group of stress proteins is the *"glucose-regulated proteins" (Grps)* which are localized in ER. Most of them are chaperones of 74 to 78 and 94 kDa. They have the homology sequences with Hsp70 and Hsp90 respectively. After stress is removed the Grps modifies into biologically inactive forms (del Razo et al., 2001). Synthesis of Grps is increased by deprivation of glucose or oxygen (van der Oost et al., 2003).

The third class of stress proteins—the stressor-specific proteins appear to participate in biochemical pathways involved in the metabolism of chemicals, metabolites or harmful intermediates that are the result of a particular chemical or physical condition rather than being a part of the cell's protective system in response to general cellular damage (van der Oost et al., 2003).

CYT P450 induction is initiated by the binding of a specific xenobiotic to a protein complex that comprises the AhR (aromatic hydrocarbon receptor) and the heat-shock protein 90 (Hsp 90). However, the cellular stress responses in fish can vary depending on tissue, Hsp family, type of stressors and fish species. It is very important to establish whether experimental procedures such as handling, sampling and other physical stressors are affecting the Hsp response because this response is very sensitive and non-specific to stressors of different nature. In order to apply the Hsp induction as a realistic biomarker in various fish species it is essential to understand the relationship between the organism stress of fish and the cellular stress response on stressor. It is very important to take into account the specificity of fish biology and ecology, geographical location, and seasonal variations in the Hsp response in the animal (Iwama et al., 2004). Chaperone level varies during life and depends on specificity of animal biology and correlates with longevity (Soti and Csermely, 2007). In experimental conditions it has been observed that very small amounts of induced Hsp can have effects on life history traits such as development, stress resistance, life span and fecundity. Hence, Hsp studies are very important for wild populations that are exposed to variable environments, including stress factors (Serensen et al., 2003). The increase of Hsp also plays an important role in enhancing the survival of the stressed fish. The stressors of different nature can increase Hsp level and wide range of abiotic and biotic stressors elicits these responses. They have genetic component but they are modified by environmental factors (Iwama et al., 1999).

Elimination of damaged molecules is essential for survival of the organisms (Grune, 2000). The important role of molecular chaperones and their universal occurrence have led the idea of a general system involved in protein quality control, operating to maintain homeostasis under normal cellular conditions. The Protein Quality System (PQS), whose major elements are sequence repair by methionine sulfoxide reductase and other enzymes, conformational repair by chaperones, protein disposal and protein sequestration into solid phase aggregates has a two-fold overall function: to secure correct folding of proteins and to assist in degradation of denatured or aggregated proteins. Variations in response of missense mutations, pathogens and other environmental stressors have been shown to correlate with specific variation in the PQS. It is an important mechanism of the resistance of the organism to unfavorable exogenous and endogenous (including genetic) factors. The cells operate to restore homeostasis disrupted by genetic stress as following: genetic stress → disrupted homeostasis → increased Hsps expression → restored homeostasis (Serensen et al., 2003). Probably all elements are present in all compartments of cells and their common features are the following: the limited overall capacity to repair/ dispose the damaged proteins, an age-dependent functional decline and an inhibition by misfolded/aggregated proteins. Damaged proteins at the sequence level may serve as chaperone substrates for conformational repair. Hsps are the part of this system and their role is increased in the stressful conditions as a result of increased concentrations of damaged proteins (Soti and Csermely, 2007).

Hsps are good biomarkers of pollution because they are part of the cellular protective response; their synthesis is to be induced by a large number of chemical, physical and biological factors; they are highly evolutionary conserved in all organisms from bacteria to plants and man. Each stress protein comprises a multigene family in which some proteins so-called cognates, constitutively expressed while others appeared in response to environmental stress (van der Oost et al., 2003). The Hsp 70 induction is more suitable as a biomarker of exposure at low temperatures. In ecotoxicological studies it is very important to take into account the specificity of fish biology and ecology, geographical location, and seasonal variations in the Hsps response in the fish.

2.4 Hematological Characteristics

Hematological parameters are also used as fish biomarkers. Their responses are non-specific to environmental stressors but they may provide important information in the ecotoxicological studies and risk assessment. Among them serum transaminases, alkaline and acid phosphatases, lactate dehydrogenase, albumin concentration and its properties, and glucose concentration are very helpful for stress responses evaluation in fish.

2.4.1 Serum transaminases

Alanine transaminase (ALT) and aspartat transaminase (AST) catalyze the interconversion of amino acids and α-ketoacids by transfer of amino groups. ALT catalyzes the transfer of the amino group from alanine to α-ketoglutarate to form glutamate and pyruvate. AST catalyzes the transfer of the amino group from aspartat to to α-ketoglutarate to form glutamate and oxaloacetate (van der Oost et al, 2003). Both aminotransferases function as a link between carbohydrate and protein metabolism. They are detected in blood serum, in the cytoplasm and mitochondria in various tissues. Enzymes are present in amphioxus of mammals, in hepatic diverticulum, in ovaries and in testes (Lun et al., 2006). Enzyme activities range widely in fish and they depend on species biological specificity (Goto et al., 2003), age (Coppo et al., 2001–2002), sex and maturation stage (Mehdi et al., 2011) and the period of reproduction (Svoboda et al., 2001). Toxicants act on the carboxyl, amino, sulfhydryl, phosphate and other groups of the molecules which result in the damage of enzyme systems by blocking active sites immobilization of essential metabolites, modification of membrane structure and its permeability. The changes of aminotransferase activities cause the disturbances of Kreb's cycle, decrease the cycle intermediates and AST and ALT compensates through providing α–glutarate (Ugwemorubong et al., 2009).

An increase of enzyme activity in the blood serum, plasma and other extracellular fluids was documented in many investigations of the organisms impacted by unfavorable conditions and it was related to organ dysfunction or internal lesions in tissues. The damaged cells release their contents (including aminotransferase) towards the blood stream and the level of these enzymes enhances in serum (Martinez-Porchas et al., 2011).

AST is used as clinic diagnostic tool and it is associated with cell necrosis of the liver and skeletal or cardiac muscle, starvation and lacking of vitamin E. Plasma ALT is an acute hepatic damage good marker (Coppo et al., 2001–2002).The increase of serum ALT activity was demonstrated in tilapia after injection of benzo[α]pyrene (BaP) (Martinez-Porchas et al., 2011). The increase of both aminotransferases activity was indicated in common carp impacted heavy metals (cadmium, lead, nickel and chromium) (Rajamanickam and Muthuswamy, 2008) and exposed in herbicide pendimethalin (Abd-Algadir et al., 2011).

We studied the effect of the fungicide cuprocsat ($CuSO_4 \cdot 3Cu(OH)_2 \cdot 1/2H_2O$) in the concentrations of copper 0.001, 0.01 and 0.1 mg l^{-1} on the activity of serum aminotransferases of Black Sea scorpion fish *Scorpaena porcus*. No differences in Cu accumulation in intact and expoed fish tissues have been observed, while enzymes activity varied and depended on the time of exposure and toxicant concentration (Fig. 2.8). Fish response displayed phase trends and the changes of ALT and AST levels were independent from each other. After the exposed fish were transferred in marine water without toxicants enzyme activities did not remain to the control level, which should explain that the liver damage caused high concentrations of Cu (Roshina and Rudneva, 2009).

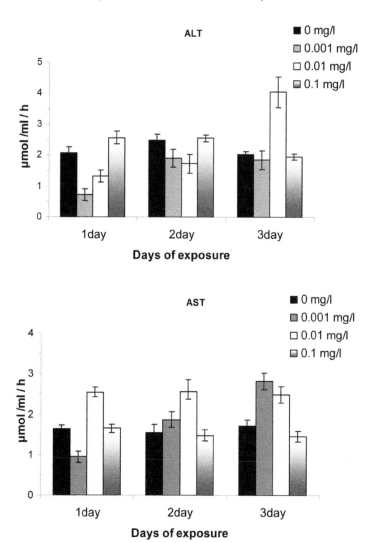

Fig. 2.8. Serum ALT and AST activity (μmol ml^{-1}h^{-1}, mean $\pm$ SEM) in blood serum of Black Sea scorpion fish *S. porcus*, exposed in different concentrations of copper (0.001–0.1 mg l^{-1}) (adopted from Roshina and Rudneva, 2009).

AST and ALT activities were elevated in blood serum of Indian Major carp *Laleo rohita* incubated in different concentrations of arsenic and chromium that indicated the hepatocytes damage (Vutukuru et al., 2007). The elevation of AST activity was observed in *Clarias gariepinus* exposed in nitrite (Ajani et al., 2011) and in crude oil (Wegwu and Omeodu, 2010) which was the result of hepatic cell membrane damage. Aminotransferase activities increased in *Oreochromis aureus* juveniles exposed to sublethal concentrations of phenol. Phenol and its derivatives changed protein metabolism by altering transamination rate of amino acids (Abdel-Hameid, 2007).

High temperature increases the toxic effects of the chemicals which were documented by the researchers. The high temperature (+30°C) increased the negative effects of organ function in tilapia after injection of BaP, which was associated with the highest aminotransferase enzyme activity (Martinez-Porchas et al., 2011). The similar effects were observed in the fish *Clarias gariepinus* exposed to different nitrite concentrations at high temperatures (Ajani et al., 2011).

However, the necrotic cells do not contribute to an increased enzyme activity in the serum and membrane defects causing increased permeability. Opposite, toxicants can also inhibit the activity or synthesis of enzymes resulting in decrease of activities in the plasma and tissues (van der Oost et al., 2003).

Most of fish diseases and pathologies might occur as a result of parasite invasion which also provokes stress in the organism. Blood parameters including aminotransferases serve as reliable indicators of fish health and resistance to infection. AST and ALT activities elevated in fish *Oreochromus nilioticus* and *Clarias gariepinus* infected the external parasites as well as urea and creatinine levels. The effect was higher in fish from heavy metals polluted locations which could be associated with the increase of enzyme synthesis in the liver of the infected animals (El-Seify et al., 2011).

The relationships between seasonal trends of trace elements concentration in tissues and aminotransferase activities in blood serum in Black Sea scorpion fish were shown in our studies (Rudneva et al., 2008a,b). Strong correlation between trace elements level in fish tissues and water temperature was noted. We observed the increase of toxicants concentration in water in warm period and decrease in cold season, which was connected with the growth of anthropogenic impact on the marine environment and the peculiarities of tested fish physiological status. The aminotransferases activity variations in different seasons were also demonstrated and the relations between trace elements concentration in tissues and enzyme activities were documented especially between ALT activity and Cu level and between AST activity and Pb, Cd and Zn concentration in fish tissues (Table 2.4).

Table 2.4. Correlation coefficients between the concentrations of trace elements in muscles and serum aminotransferase activities in Black Sea scorpion fish *Scorpaena porcus* (Rudneva et al., 2008a).

Enzymes	Cu	Pb	Cd	Zn	As	Hg
ALT	0.766	0.179	0.381	0.201	0.059	−0.32
AST	0.218	0.995	0.727	0.847	−0.280	−0.53
Ratio ALT/ AST	0.302	0.964	0.176	0.995	0.127	−0.067

In our studies we noted the decrease of serum aminotransferase activities in *S. porcus* caught in polluted site (Fig. 2.9).

In our study aminotransferase activities (both ALT and AST) were lower in serum of fish from polluted sites (the Streletskaya Bay and the Karantinnaya Bay) than those in the samples collected in free polluted area. We propose that the reason of this phenomena was the lack of serum protein concentration in fish from contaminated locations which was 1.5-2-fold lower than that in fish from reference site. This explanation is confirmed with high correlations between protein concentration in serum and ALT and AST activities (r = 0.79 and r = 0.98 respectively). On the other hand, aminotransferases are the organ-specific indicators for toxic effects and determination of transaminases in serum has been proved in diagnosis of liver damage (Van der Oost et al., 2003).

No statistically significant alterations were shown in ALT and AST activities in blood plasma of flounder collected in different contaminated sites of the Baltic Sea (Napierska et al., 2009). Thus, the response of serum transaminases is not uniform and could be used as indicator of fish physiological status and health. A significant increase in ALT activity was

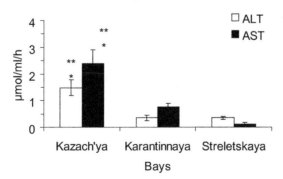

Fig. 2.9. Aminotransferases activity (µmol per ml per h, mean ± SEM) in fish serum. Asterisk *—significant (p < 0.05) difference with the values of the Streletskaya Bay, **—significant (p < 0.05) difference with the Karantinnaya Bay (Rudneva et al., 2012).

observed in three laboratory studies (60%) and none of the field studies, while a strong increase (> 500% of the control) was not observed in any of the studies considered. The AST responses for all fish species from eight laboratory studies and seven field experiments demonstrated a significant increase in AST activity in 38% of the laboratory studies and in two field studies (29%), while a strong increase (> 500% of the control) was only observed in one laboratory study (13%) and in one field study (14%). The authors have concluded that ALT and AST activities in blood plasma have been applied in fish to indicate bacterial, viral and parasite invasion, intoxication under water pollution but they cannot be considered reliable biomarkers to evaluate the effects of chronic pollution exposure in monitoring studies and applying in ERA (van der Oost et al., 2003).

2.4.2 Serum enzymes and other parameters

Lactate dehydrogenase (LDH) as metabolic enzyme is useful indicator in aquatic toxicology for the evaluation of stress in fish. LDH level in serum of carp exposed in heavy metals was significantly decreased on the first day and then progressively increased in the following 32 days of the experiment (Rajamanickam and Muthuswamy, 2008). LDH activity elevated in *Oreochromis aureus* juveniles exposed to sublethal concentrations of phenol which was explained by the alterations in carbohydrate metabolism (Abdel-Hameid, 2007).

Alkaline phosphatase (ALP) is membrane-associated glycoprotein which is produced in the liver and plays a role in transport of ions and absorption of water across cell membranes. It is used as diagnostic tool for evaluation of liver function and its pathologies and its activity depends on fish age, maturation stage, water pollution, food composition, temperature and peculiarities of fish biology and ecology (Mehdi et al., 2011). Significant growth of serum ALP is an indicator of cholestasis and secretory function of the hepatocytes (Ugwemorubong et al., 2009).

We observed seasonal variations of aldolase activity in serum of scorpion fish *S.porcus* and high positive correlation between Cd concentration in tissues and enzyme activity ($r = 0.862$) while between aldolase activity and Cu and Zn level the link was negative ($r = 0.925$ and $r = -0.702$ respectively) (Rudneva et al., 2008a).

Lysozyme plays an important role in non-specific immune response and it has been found in various tissues in fish. Lysozyme protects the organism against parasite and bacterial infections and its levels varied in fish depending on natural conditions and individual animal resistance to environmental stressors (Mehdi et al., 2011).

Plasma cortisol decreased in fish injected benzo[α]pyrene (BaP) and it was greater in higher temperatures, glucose was increased in contaminated fish (Martinez-Porchas et al., 2011). Similar effect was observed in fish *Clarias gariepinus* exposed to different concentrations of crude oil. Toxicants may have negatively affected the biochemical pathways that is responsible for the balance of body energetic sources including glucose content (Wegwu and Omeodu, 2010).

2.4.3 Serum proteins

Serum proteins play an important role in transport of different substances, defense of the organism against pathological agents and some other functions. Our studies of serum proteins composition of Black Sea scorpion fish *S. porcus* showed the differences between fish caught in polluted and non-polluted locations (Rudneva et al., 2005). In fish from polluted area the electrophoretic spectra of serum proteins was more heterogeneous and the number of bands varied wide than in fish from non-polluted region. Electrophoretic characteristics and distribution of protein bands, especially albumin also differed in fish from both environments. Among serum proteins albumin is the major protein of the blood which plays an important role in transport of wide range of physiological and exogenous ligands(drugs) and endogenous chemicals (fatty acids, hormones, bilirubin) regulation of the colloid osmotic pressure of the blood (De Smet et al., 1998; Baker, 2002).

Albumin-like proteins were found in different bony fish and lamprey, but in elasmobranchs albumin was absent in some species (Metcalf and Gemmell, 2005). The information of albumin presence in teleosts was also contradicted. Concentration of albumin-like proteins in fish plasma of teleosts can vary from 10% to 50% while in terrestrial vertebrates albumin accounts for more than 50% of the total serum proteins concentration (McDonald and Milligan, 1992). Besides that, fish physiological status and environmental factors impact albumin and may change its characteristics: the maturity process of gonads and maturation stage of fish (Ishioka and Fushimi, 1975), seasonal variations (Schlotfeldt, 1975), pathological processes and parasites invasion (Moeyner, 1993), sex (Zowail et al., 1994), chemical toxicants (Richmonds, 1990) and long-term pollution (Moussa et al., 1994; Sayed et al., 2011).

Our findings indicate that fish physiological status, age, season and habitats play an important role in serum protein properties, especially albumin. The alterations of albumin electrophoretic mobility connected for the first time with its transport function and the differences between the round goby from Black Sea and Azov Sea could be the result of genetic mutation and the specificity of environmental factors, including pollution (Rudneva and Kovyrshina, 2011). We also reported that in pollution sites protein and albumin concentration in fish blood serum were significantly lower as compared to fish characteristics from reference areas (Rudneva et al., 2012) (Fig. 2.10).

Therefore, enzyme (ALT, AST, LDH, ALP, etc.) activities in blood plasma, serum proteins level and composition have been applied in fish to indicate bacterial, viral and parasite invasion, intoxication under water pollution but they cannot be considered reliable biomarkers to evaluate the effects of chronic pollution exposure in monitoring studies because they are not specific and could be used as a tool for diagnostis of fish status under the impact of majority of abiotic, biotic and anthropogenic factors and their combination.

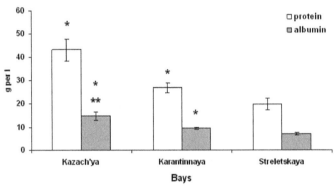

Fig. 2.10. Total protein and albumin concentrations (mean $\pm$ SEM) in fish serum. Asterisk *—significant ($p < 0.05$) difference with the values of the Streletskaya Bay, **—significant ($p < 0.05$) difference with the Karantinnaya Bay.

2.5 Metabolic Rate

An energetic approach of stress reactions is due to its reliance on first principle of thermodynamics that matter and energy are conserved. If an additional cost like damage from stressors exposure is imposed upon it, then fish must reallocate energy expenditures in order to repair the impacted system. Reallocation will most likely be reflected by reduced growth, development or reproduction because energetic expenditures are distributed in all levels of biological organization and they play an important role in animals life. Another advantage of an energetic approach is that energy is the common currency of ecology. It provides a means for quantifying the value of available resources as well as associated acquisition costs and can account for physiological effects of suboptimal physiological conditions that may deplete energy reserves (Beyers et al., 1999).

Basic energy budget equations balance is as follows:

$$C = R + A + S + F + U + \Delta B$$

where C is food consumption; R is metabolism; A is activity, S is specific dynamic action (the cost of processing food), F is ejection (feces), U is excretion (urine) and ΔB is somatic growth and reproductive growth in mature animals.

Growth represents a cumulative integrated response to the complex of environmental and physiological variables that influence fish. Physiological stress could be quantified in terms of energy by measuring routine metabolism, food consumption, activity and growth rates. Metabolic rate is a good measure of the energy being expended for compensative mechanisms at stress conditions because it integrates all physiological processes (Beyers et al., 1999).

Studies of the metabolic responses of marine organisms to stressors are determined by various analytical techniques, including respirometry, biochemical and physiological analysis and direct calorimetry (Keckeis and Schimer, 1990; Canepa et al., 1997; Parra and Yufera 2001). All biological processes result in either heat production or heat consumption and one of the versatile technique for a non-destructive and non-invasive studies of the animals to measure total metabolism, independent of oxygen consumption is microcalorimetry (Baker and Mann, 1991; Russel et al., 2009). Besides that microcalorimetry is used for *in vivo* determinations and it's a great advantage as compared with the other biochemical and physiological methods. At present a lot of methods are used for biomarkers determinations. One of them is microcalorimetry which is possible to assay the metabolic rate in relation to activity organism and to estimate the energy costs at the normal and toxic conditions (Handy and Depledge, 1999).

Heat production of different species varies as a consequence of both exogenous (temperature, salinity, oxygen consumption, pressure, feeding, pollution, etc.) and endogenous factors (animal size, growth, age, behavior, activity, physiological status and biochemical composition) (Haegen et al., 1994; Lamprecht, 1998; Normat et al., 1998; Maenpaa et al., 2004). The metabolic rate of the organism changes under the impact of many environmental factors including physical, chemical and biological (infection) and their toxic effects may also be measured by microcalorimetry also.

The relationship between pollutant exposure and metabolic rate of the test-organisms is well documented. But the trends of the heat production in toxicants-treated organisms are varied widely and in some cases the authors obtained contradictory results (Penttinen and Kukkonen, 2006). Metabolic costs are associated with maintenance/protection of the organism against environmental stressors (Penttinen et al., 2005; Penttinen and Kukkonen, 2006; Penttinen and Holopainen, 1995). At the same time metabolic rate can vary considerably with causes other than stressors. It tends to fluctuate due to many organism-specific factors such as growth, developmental stage, maturation, spawning, etc. Thus it is very important to understand the mechanism of organism-specific variability in terms of energetic responses to environmental factors including pollutants.

The metabolic response should be separate through monitoring and the mechanisms of toxicity could be different and dependent on the chemical structure and origin of xenobiotics and on tested organisms (Penttinen and Kukkonen, 1998; Penttinen et al., 2005). Toxicity reduces metabolic rate which was demonstrated in moving activity, decrease of enzyme activities, blood vessel pulse in some invertebrates (Maenpaa et al., 2004; 2009). The ability to reduce metabolic rate during the exposure to environmental stress, termed metabolic rate suppression is an important component to enhance survival in many organisms. Metabolic rate suppression can be achieved through modifications to behavior, physiology, and cellular biochemistry, all of which act to reduce whole organisms energy expenditure (Richards, 2010).

Some environmental stressors affect via a specific mechanism of toxicity such as uncoupling of oxidative phosphorilation, thus affecting the regulation of energy generation and dissipation and direct measurements of the rate of energy metabolism demonstrated the increase metabolic rate of the animal impacted unfavorable factors (Widdows and Donkin, 1991). Besides that thyroid function may represent a sensitive target for environmental estrogenic compounds and the disturbance of endocrine metabolism may modify energetic metabolism also (Penttinen et al., 2005). Penttinen and Kukkonen (2006) showed significant increase of heat production of the alevins of *Salmo salar* exposed in PCP in different concentrations. The output-mediated link between tissue residues concentration and heat output was demonstrated and the high correlation between bioaccumulation of PCP and ecotoxicity was revealed.

Change of heat dissipation under the impact of phenol chemicals (2,4-DNP, PCP, 2,4,5-TCB) is also demonstrated on aquatic invertebrates oligochaete *Lumbriculus variengatus*. The changes of heat output reflected the relationships between chemical concentrations and metabolic rate. Increase of metabolic rate was closely correlated with the dose of toxicants (Penttinen and Kukkonen, 1998). At the same time the effect of toxicants on metabolic rate differed among tested animals belonging to various taxonomic groups which demonstrate the different sensitivity of the animals to toxicants (Penttinen et al., 1996).

It is well-documented that exposure to metals led to such negative effects on aquatic animals including fish and invertebrates, which associated with the damage of respiratory, swimming behavior, histopathology, accumulation in body tissues and damage of biochemical pathways. Several investigators documented an increase of energy requirements to resist or repair toxicant-induced damage in aquatic animals. In contrast some metals are known to either decrease or do not change the metabolic rate. The reasons of the metabolism modification in aquatic animals exposed in metals could be explained with the hypometabolism likely to stem from damage of gill epithelium, neutoxicity or hypoactivity and these three factors that can also ultimately affect growth, survival and reproduction. At the same time despite the high concentrations of Hg accumulated by mosquitofish in the experimental conditions no difference in metabolic rate among individual from three populations was found (Hopkins et al., 2003). In contrast acute exposure to dissolved inorganic Hg increased metabolic rate of mosquitofish from the same source population (Tatara et al., 1999). The authors suggested that chronic environmental contamination led to the general resistance of fish to Hg and subsequent attenuation of effects on metabolic rate (Hopkins et al., 2003).

Chronic impact of environmental pollution on fish energetic metabolism was also documented also (Hopkins et al., 2003; Pane et al., 2003; Rudneva et al., 1998; Rudneva and Shaida, 2012). Our findings demonstrated that the impact of different chemical stressors (pesticides, PCB, environmental pollution) modified metabolic rate of fish and resulted in the disturbance of the processes of energy generation and utilization (Rudneva and Shaida, 2012). Thus many environmental factors like biotic, abiotic and anthropogenic may modify fish metabolism and application of microcalorimetry along with other analytical methods should also be helpful for the explanation of the mechanisms of developmental changes in fish, the influence of xenobiotics disturbance and interactions between the organism and environment.

The changes in metabolic rate is associated with the secondary response on stress which comprises the various biochemical and physiological alterations and is mediated to some extent by hormones. Adrenaline and cortisol activate a number of metabolic reactions that result into the alterations in blood chemistry. The response of the organism on stress is an energy-demanding process and the animal mobilizes energy substrates to cope with stress metabolically (Iwama et al., 1999). Many fish are able to survive in polluted environments but tissue levels of toxicants are often maintained at levels below of hose observed in the environment. It has been suggested that this fact is mediated by the multixenobiotic resistance (MXR) phenomenon. The MXR mechanism acts as an energy-dependent pump that removes both endogenous and exogenous chemicals from the cell, thus preventing their accumulation and cytotoxic and the protein responsible for this transport function is the transmembrane P-glycoprotein (PGP) (van der Oost et al., 2003).

Metabolic stress is associated with the depletion of cellular ATP because the energetic costs of response of the organism increase caused the induction of various components of cell defense (Hsps, antioxidant enzymes, MFO, etc.) (Widdows and Donkin, 1991). Transcription and translation are energetically expensive processes and, for instance, in fish protein synthesis is estimated to cost 40% of the total oxygen consumption (Lyndon et al., 1992).

Several researchers have shown a correlation between decrease in cellular energy (ATP) and Hsp activation. Changes of ATP level have been potential to affect several aspects of Hsp function and expression. ATP depletion is thought to cause accumulation of denaturated proteins and aggregation of constitutive Hsp70 that ultimately triggers the heat shock response under non-heat temperature. It has been proposed that during metabolic stress (i.e., decrease in ATP) the constitutive and inducible forms of Hsp70 become stable bound with proteins that are unable to fold or have become denaturated. As the available Hsp70 concentrations decrease, the cells respond by increase Hsp expression through a feedback mechanism. At the same time heat shock alone had no significant effect on ATP level but induced increased gene transcription of the major stress protein Hsp70 in rainbow trout blood cells (Currie et al., 1999).

At high temperatures a significant decline in the ratio of ATP/ADP was observed. This decrease is indicative of an elevation in ADP that accelerates energy turnover and equilibrates ATP production with ATP consumption. In this case a decrease in ATP/ADP signals an energetic deficiency and results in regulatory response. Any alterations in the adenylate ration indicates that metabolism is changing, regardless of ATP concentration. Hypoxia causes a greater metabolic disturbance than heat shock alone, as indicated by the decrease in ATP/ADP (Currie et al., 1999).

Glucose is continuously required as energy source by cells and tissues. Glucose level in blood plasma is maintained in general through the conversion of hepatic glycogen. It depends on fish physiological status, season, maturation stage, sex and environmental factors such as food consumption, water temperature, etc. Increase of plasma glucose concentration may be due to acceleration of glucose production or release. Glucose levels vary widely in various species and have shown to increase at stress conditions (Mehdi et al., 2011). The production of glucose with stress assists the organism by providing energy substrates to tissues such as brain, gills, and muscle, in order to cope with increased energy demand. The stress hormone adrenaline and cortisol have been shown to increase glucose production in fish by both glucogenolysis, and elevation glucose level in plasma (Iwama et al., 1999). For instance, glucose content in the liver, gills and muscle of blue tilapia *Oreochromis aureus* juveniles generally reduced due to phenol exposure and the changes were concentration-dependent. The reduction of hepatic glycogen storage could be possibl because the hepatic synthesis of detoxifying enzymes requires high energy levels. In addition, the total protein content was decreased in the liver and gills in the animals exposed in phenol. The researchers suggested that protein was taken as an alternative source of energy for detoxification processes (Abdel-Hameid, 2007).

Metabolic rate, consumption and growth were influenced by chemical exposure and energetic costs of contaminated-induced alterations in metabolism and food consumption in fish can be integrated with bioenergetics model to demonstrate biological effects of chemical pollution of wild fish populations (Beyers et al., 1999). Complex physiological traits such as routine aerobic metabolic rate or exercise performance, are indicators of the functional integrity of fish that can reveal sub-lethal toxicological effects of pollutants (McKenzie et al., 2007). The authors have shown that swimming performance and metabolic rate are a good biomarkers of pollutant exposure because fish collected in European rivers with different level of pollution show the ranges of metabolic characteristics. They directly reflect physiological status of fish and they are also of immediate relevance to fish ecology. Measurements of oxygen uptake during swimming test revealed increased rate of routine aerobic metabolism in both chub and carp at polluted sites in all the rivers examined indicating sub-lethal metabolic loading effect of exposure to complex mixtures of chemicals. However, they may be less specific that biochemical or molecular biomarkers in terms of revealing particular types of pollution. They are reliable integrated measures of the responses of many physiological systems and so can provide insight into why fish fail to colonize some polluted inhabitants (McKenzie et al., 2007).

High concentration of CO_2 in marine ecosystems caused algae bloom or hypoxic conditions in deep water layers provokes stress in fish. The study of correlated changes in energy and protein metabolism in two arctic fish species at various levels of pCO_2 of intra- and extracellular pH has been shown. A decrease in extracellular pH from control levels (pH 7.90) caused a reduction in aerobic metabolic rate of 34–37% under both normocapnic and hypercapnic conditions. Protein biosynthesis was inhibited by about 80% under conditions of severe acidosis in hepatocytes from both fish species. Elevation of pCO_2 may limit the functional integrity of the fish liver due to a pronounced depression in protein anabolism which may contribute to the limits of whole-animal tolerance to raised CO_2 levels (Langenbuch and Portner, 2003).

Hence, metabolic rate of fish is a very sensitive parameter for determination of water toxicity. It could be applied as biomarker to the evaluation of water quality and it has taken an effective role in solving many environmental problems such as the interactions between environmental factors and fish resistance, toxic effects and fish protection.

Sensitive endpoints should be used in risk assessment because already low environmental concentrations of toxicants may cause damage of fish inhabited impacted locations.

2.6 Cholinesterases

Enzymes cholinesterases (ChE) may reflect the disturbance of neural functions of the organism in stress conditions. Two types of ChE are presented: one with a high affinity for acethylcholin (AChE) which is synthesized in hematopoiesis occurs in the brain, endplate of skeletal muscle, erythrocyte membranes and its main function is to regulate neuronal communication by hydrolyzing the ubiquitous neurotransmitter acetylcholine in synaptic cleft, and those other with a high affinity for butyrylcholin (BChE) also known as non-specific esterases or pseudocholinesterases. The second is synthesized in the liver and it is present in plasma, smooth muscle, pancreas, adipocytes, skin, brain and heart. BChE is pointed as one of the main detoxifying enzymes able to hydrolyze or scavenge many kinds of xenobiotics, including pesticides. One of the possible functions of BChE is to protect AChE against anticholinesterastic agents. ChE may play other roles in the neuronal tissues, such as differentiation and development, adhesion and signaling. AChE controls ionic currents in excitable membranes and plays an essential role in nerve conduction processes at the neuromuscular junction (Sarkar et al., 2006). AChE also participates in hematopoietic differentiation. AChE class is more homogeneous of the primary structure than the class of BChE (Assis et al., 2011).

Fish brain contains AChE, but not BChE, while muscle contains both AChE and BChE. The study of muscle tissues from 28 fish species observed that AChE/BChE ratio ranged from 0.5 to 3.3 and these cholinesterases are extremely sensitive to organophosphate (OPs) and carbamate (CBs) pesticides (Magnotti et al., 1994). OPs and CBs act by phosphorilating or carbamoylating the serine residue at the active site of the ChEs. Their structure presents either similarities to the substrates or their hydrolytic intermediates and interact very slowly with the enzyme by forming stable conjugates. This mechanism hinders the normal functioning of the enzyme which cannot prevent the accumulation of the neurotransmitter in the synaptic cleft. The overstimulation caused by acetylcholine continuously firing its receptors generates a range of signs and symptoms (Assis et al., 2011). Apart from the insecticides, a few other contaminants, including heavy metals (Cd, Hg and Cu) were found to show anticholinesterase activity (Sarkar et al., 2006). In addition, AChE inhibition in fish has been recorded following exposure to pyrethroids and cyclodiene pesticide groups.

A variety of studies with estuarine fish species have suggested that brain AChE inhibition levels of > 70% are associated with mortality in most species, but some animals appeared capable of tolerating much higher levels (> 90%) of brain inhibition, the other estuarine fish sublethal effects were associated with brain AChE inhibition levels as low as 50% (Fulton and Key et al., 2001). The slow interaction between enzyme and pesticides is behind the ability ChE has to signal inhibition several days or weeks after exposure, even when the concentration in the water is negligible. On the other hand it is possible to gain more precision in correlation between pesticide levels and the resulting inhibition (Assis et al., 2011).

Monitoring of AChE inhibition in the tissues of marine fish has been done as a good tool for assessing pesticide pollution of the marine environment (Kirby et al., 2000). Due to interindividual variability an AChE inhibition of > 20% is often required to confirm the presence of exposure. The variability of AChE inhibition in the organisms depends on animal age, size, season, water temperature and pH and demonstrates clear interspecies differences which may be a function of fish's general activity and motorability (Georgiades and Holdway, 2007). ChE inhibition by OP compounds follows different behaviors depending on pesticide chemical structure. Ions can alter ChE activity inhibiting or activating so it was proposed that enzymes could be used as biomarkers of heavy metals pollution. Another concern about application fish ChEs as biomarker of OPs and CBs pesticides is that some species of cyanobacteria produce anticholinesterasic metabolites such as anatoxin-a which can be considered as natural OP compounds in which toxicity can be approximately 1000-fold higher than that of the insecticide paraoxon (Assis et al., 2011).

AChE is involved in the hydrolysis of the neurotransmitter acetylcholine at nerve endings to terminate nervous stimulation, preventing continuous nerve firings which is vital for normal functioning of sensory neuromuscular systems. AChE inhibition disables the main mechanism that terminates cholinergic synaptic transmission in the brain and neuromuscular junctions of both vertebrates and invertebrates. The inhibition of the enzyme can have biological consequences such as hyperactivity, tetanus and death (Tierney et al., 2007; Georgiades and Holdway, 2007). Many kinds of organophosphate and carbamate pesticides are documented as effective AChE inhibitors. Hence AChE inhibition has been used as biomarker of agricultural impact on natural populations including fish. However, complex mixtures of pollutants, other than pesticides, could be important sources of AChE-inhibiting compounds in the aquatic ecosystems and additional research is needed to understand the toxic mechanisms and the peculiarities of species response (van der Oost et al., 2003).

The target enzyme of the organophosphorus pesticides or their metabolites is assumed to be AChE. Enzyme activities in brain and muscle were quickly inhibited during a 48-hr *in vivo* exposure to insecticides-chlorpyrifos (0.1 ppm), parathion (0.15 ppm), and methyl parathion (8 ppm) in mosquitofish (*Gambusia affinis*). The recovery of the enzyme activities demonstrated very slow rate in fish and remained persistently inhibited and/or they were not being replaced at any significant rate after exposure in pesticides (Boone and Chambers, 1995). Similar effects have been shown in mussels which were exposed to 0.1 mg l⁻¹ organophosphate pesticide azamethiphos for periods of up to 24 h. A significant reduction in acetylcholinesterase activity in hemolimph and in the gill (Canty et al., 2007)was noticed. Our study also demonstrated that serum ChE activity was the lowest in scorpion fish from high polluted area (Fig. 2.11). Decrease of enzyme activity may reflect the pollution of organophosphate chemicals, metals, detergents which were present in domestic sewage and maritime effluents entering into the Streletskaya Bay (Rudneva et al., 2012).

The inhibition of AChE activity has often been associated with swimming activity of fish, but the results obtained by researchers are contradictory. The authors documented an increased swimming activity in goldfish exposed to carbofuran following 4 h (Bretaud et al., 2001), while coho salmon exposed to chlorpyrisos for 96 h did not display hyperactivity. The concentration-dependent decrease in swimming activity was correlated with the trends of AChE activity. Significant decrease in swimming activity was seen in cases of exposure to concentration as low as 0.6 µg l⁻¹ which corresponded to a relative AChE activity of 77% of the control. The authors reported that the response of the fish AChE and swimming activity depended on pesticide concentration (Tierney et al., 2007).

Inhibition of AChE was accompanied by an increase in acetylcholine levels which can result in the increase of catecholamines concentration that can affect the induction of enzyme of glycogenolysis and glycogen synthesis. Continuous stress may affect the synthesis site of AChE or decrease the levels of excess AChE. Fish mortality may be due to inhibition of other enzymes especially involved in carbohydrate and protein metabolic pathways. The inhibition of AChE activity might have indirect influence of energy metabolism in nerve and brain cells (Adedeji, 2011).

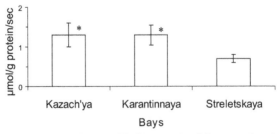

Fig. 2.11. Cholinesterase activity (mean ± SEM) in scorpion fish serum. Asterisk *—significant (p < 0.05) difference with the values of the Streletskaya Bay.

There are two main reasons to apply fish ChE as biomarker of water pollution. The first concerns the availability of this sources: in 2009 the world fisheries and aquaculture production was 145.1 million tones, and most of the fish waste reused comes from tissues other than those that provide ChEs. Very high AChE concentrations were found in electric organs of fish. According to the FAO, 20% inhibition of brain AChE activity is considered the endpoint to identify the no-observed-adverse-effect-level (NOAEL) in organism, while signs and symptoms appear when AChE is inhibited by 50% or more. Death occurs above 90% inhibition (Assis et al., 2011).

Therefore, monitoring of AChE inhibition in the tissues of marine fish has been considered as a good tool for assessing pesticides and heavy metals pollution of the marine environment. The variability of AChE inhibition in the organisms depends on animal age, size, season, maturation, water temperature and pH and demonstrates clear interspecies differences which may be a function of fish general activity and motorability. Many kinds of organophosphate and carbamate pesticides are documented as effective AChE inhibitors. Hence AChE inhibition is used as biomarker of agricultural impact on natural populations including fish. However, complex mixtures of pollutants, other than pesticides, could be important sources of AChE-inhibiting compounds in the aquatic ecosystems.

2.7 Endocrine Disruption

Many xenobiotics produced over the last century have the potential to affect the vertebrate neuroendocrine system, which regulates vital processes including development, growth, metabolism and reproduction. Chemicals that either mimic or antagonize the actions of naturally occurring hormones are termed endocrine disruptors or endocrine disrupting chemicals (EDCs). These include estrogenic EDCs (environmental estrogenes), substances that elicit an estrogenic response by mimicking the action of endogenous estradial-17β (E$_2$). Examples of estrogenic EDCs include biodegradation products of alkylphenol polyethoxylates, polychlorinated biphenils (PCBs) and pesticides, as well as synthetic estrogenes as athinylestradioal (EE$_2$) and diethylstilbestrol (DES). Many EDs affect reproductive function, because of the lipophilic and persistent nature they accumulate and biomagnify in aquatic biota (Porte et al., 2006). We studied the effects of PCB on some biochemical characteristics of Black Sea fish species *Scorpaena porcus* and *Mullus barbatus ponticus* and the highest concentration of examined xenobiotics were indicated in fish gonads and liver (Rudneva, Zherko, 1999, 2000). Xenobiotics may cause lesions, hemorrhage, or malformations in the gonads, pituitary, liver and the brain. Production and secretion of the hormones of the hypothalamus, pituitary and gonads is usually inhibited and their metabolism in the liver can be altered (van der Oost et al., 2003).

The measurements of yolk precursor protein, vitellogenin (Vtg) have been used frequently for the assessment in bioassays of animal exposure to estrogenic EDs (Eidem et al., 2006). Vitellogenin is a species-specific and female-specific metal-binding (Ca, Zn, Cd, Fe, etc.) lipoglicophosphoprotein with molecular masses ranging from 300 kDa to 600 kDa synthesized by hepatocytes, released into the bloodstream and actively sequestered by maturing oocytes as precursors to yolk proteins induced by estrogens (Nicolas, 1999; Hiramatsu et al., 2006) which were detected in serum or plasma. The general process of vitellogenesis in fish in details was observed in many publications and it has been summarized in the review of Hiramatsu et al. (2006) who described the multiplicity of vitellogenin in various fish species and its gene groups.

The following characteristics make Vtg a functional biomarker for exposure of fish to estrogenic EDCs in the aquatic ecosystems (van der Oost et al., 2003; Hiramatsu et al., 2006):

- fish species have been used extensively in field studies of EDCs and water quality;
- vitellogenesis (oocyte maturation and yolk incorporation) is a hormonally controlled and regulated process. In fish, the primarily exogenous synthesis of vitellogenin is initiated by gonadotropins and regulated by estrogens. In polluted areas its induction is a specific physiological response of fish to estrogen or estrogenic chemicals;
- induction of Vtg synthesis by estrogen (estradiol) is dose-dependent process within broad limits;
- Vtg appears naturally in maturing females but not in males or immature fish although vitellogenesis is induced in males and juveniles exposed to estrogen or estrogenic EDCs;
- the presence of Vtg in the blood or Vtg transcripts in the liver of male or juvenile fish may be taken to indicate past or current exposure to estrogen or estrogenic EDCs.

Another potential biomarker of endocrine disruption in male fish might be the induction of Zona radiata protein (Zrp) also known as Vitelline envelope protein (Vep) which is produced during zonagenesis. Zonagenesis is the process of synthesis of eggshell zona radiata proteins, their transport and deposition by the maturing oocyte. Zrps are glycoproteins that form the chorion of the developing egg. Three such proteins (Vep 1, 2 and 3) were identified in fish, they are normally synthesized in females only during oogenesis starting before the onset of vitellogenesis. Male fish may also produce Veps following exposure to xenoestrogens. However, Veps are hydrophobic proteins and they are difficult to measure using conventional methods while Vep mRNA can be assayed with relative ease for the cloned sequence. Vtg and Vep are not mediated via same mechanism because Vep induction can be regulated by cortisol in addition to estrogens (Arukwe and Goksoyr, 2003; Hutchinson et al., 2000, 2006).

Both Zrp and Vtg are synthesized in the liver under endocrine regulation through the hypothalamic-pituitary gonadal-liver axis. Recent studies with juvenile salmon indicated that the Zrp response was more sensitive to various environmental pollutants than the Vtg (Arukwe and Goksoyr, 2003).

The indirect methods could be used to study the alterations provoked by EDC such as increase in RNA contents, lipid deposition, glycogen depletion, increase in protein levels, calcium, magnesium, zona-radiata proteins (Zrp) and phosphoproteins contents by the alkali-labile phosphate (ALP) method. In fish, ALP levels have been shown to associate with Vtg levels. The dramatic induction of Vtg and Zrp in adult male swordfish was correlated with the PCB concentrations in fish tissues (Fossi et al., 2002; Porte et al., 2006). The result obtained by several investigators confirms that induction of Vtg and Zrp can be used as diagnostic and prognostic tool for exposure assessment of fish to EDCs in environment.

There are some androgen biomarkers which focused on characteristics of male gender development that are controlled by androgens. These parameters include the facial tubercles and nuptial pads in fathead minnow and induction of spiggin in sticklebacks. These biomarkers have shown to be highly sensitive to a range of environmental androgens and antiandrogens (Hutchinson et al., 2006).

The mechanism of action of EDCs can be divided into agonistic/antagonistic effect ("hormone mimics"), disruption of production, transport, metabolism, or secretion of natural hormones disruption of production and/or function of hormone receptors. The number of nuclear hormone receptors being potential targets for EDCs has increased dramatically in the last decade. Among them there are the compounds indicating that one single compound exerts all the three mechanisms, depending on the dose given to the organism. The researchers explain this phenomena that the nature of the effect of an EDC is determined by dose-dependent routing and cross-talk between different classes of nuclear receptors (Goksoyr, 2006).

The relationship between elevated serum Vtg and gonadal abnormality and physiological characteristics such as gonad morphology, sperm motility, fertility, and sex differentiation in fish was documented. The appearance of low levels of Vtg in the serum of naturally maturing males, males fed with diet containing phytoestrogens, and males exposed to estradiol excreted by females living in close proximity has been observed in a variety of species.

As a practical result the researchers determined a threshold for normal baseline levels of Vtg to be 10µg Vtg per ml in the serum of wild male salmonids based on annual changes in Vtg levels in cultured, reproductively normal male fish. The typical baseline level of Vtg in fish plasma ranges and depends on fish species biology. In addition hepatic expression of Vtg transcripts also can be used to evaluate animal exposure to estrogenic EDCs and potentially may offer greater sensitivity than measurements of Vtg protein, but interpretation of Vtg gene expression might be entirely different from that of Vtg protein levels (Hiramatsu et al., 2006).

A large amount of research has been directed at assessing the effects of polycyclic aromatic hydrocarbons (PAHs) on different phases of vitellogenesis in fish and identifying their mechanisms of action. PAHs have been found to have a deleterious effect on the vitellogenesis of fish from feral populations as well as in laboratory experiments. This is of particular concern as any of the stages of the vitellogenic cycle could be affected which would result in harmful sub-lethal effects with trans-generational consequences. The reported effects include reduction in circulating hormones and plasma vitellogenin, estrogenic and antiestrogenic effects, retardation of oocyte maturation and reduction of reproductive success and appearence of intersex gonads. However, the lack of agreement between some of the results indicates that some of the mechanisms are not yet completely understood and require further investigation (Nicolas, 1999). In addition, CYP1A activity was inversely related to the egg viability, the fertilization and development success. The responses in CYP1A activity interfered with the ability of the general class of CYT P450 enzymes to regulate sex steroids. Due to their broad substrate specificity, isoenzymes may accelerate testosterone clearance rates and affect estradiol levels in contaminant-exposed fish. Binding of xeno-estrogenic xenobiotics to the estrogen receptor also causes estrogen effects (van der Oost et al., 2003). In addition, phenolic derivates are a group of EDCs and they can also induce sex reversal and it may be possible that phenol stimulates Vtg synthesis in fish liver (Abdel-Hameid, 2007).

The fauna of indoor seas such as Mediterranean Sea, Black Sea and Azove Sea whose basins have limited exchange of water with oceans and are surrounded by some of the most heavily populated and industrialized and agricultural countries ("hot ecolohical points") could be a target for EDCs. In these environments top predators such as pelagic fish and marine mammals tend to accumulate large quantities of toxicants and these species are potentially "at risk" due to EDCs contamination (Porte et al., 2006). Decreased reproductive capability in feral organisms may be considered as one of the most damaging effects of persistent pollutants which have endocrine activity. Low-level pollution may decrease the fecundity of fish populations, leading to a long-term decline of biodiversity and eventually extinction of important natural resources (van der Oost et al., 2003).

Fish biomarkers of exposure (Vtg and Zrp induction in male fish) are powerful tools for endocrine disruptors distribution in the aquatic environment. Recommendations from advisory board to US Environmental Protection Agency include expansion of the scope of the program (Hutchinson et al., 2006):

- to include not just estrogens but also chemicals that affect any aspect of the hypothalamic-pituitary-gonadal (HPC) and thyroid axes;
- consider wildlife, in addition to human health effects;
- expand the chemical universe of concern to all chemicals on the US EPA TSCA (Toxic Substances Control Act) inventory (> 80,000 compounds).

For practical reasons the fish species which were included in OECD (Organization for Economic Co-operation and Development) endocrine disruptors monitoring program are: fathead minnow (*Pimephales promelas*), medaka (*Oryzias latipes*), rainbow trout (*Oncorhynchus mykiss*), sheeps head minnow (*Cyprinodon variegates*), and zebrafish (*Danio rerio*). Fish screening assays include fish gonadal recrudescence assay (GRA) which was provided in fathead minnow of both sex in "winter phase" and in "summer phase" conditions. The evaluation of secondary sexual characteristics, gamet quality, fecundity, pathology of gonads, measurement of relative gonad weight and Vtg levels are used as indicators of endocrine disruption. An alternative approach is juvenile fish screening assay, because juvenile animals are very sensitive to estrogens and antiestrogens and it is possible to use them at specific endpoints (Vtg and Zrp levels) (Hutchinson et al., 2000).

Therefore, the result obtained by many investigators confirm that induction of Vtg and Zrp in fish from the polluted sites may be used as diagnostic and prognostic tools for exposure assessment of fish to EDs in environments. Vtg levels and EDC response range between various species including fish and depend on several reasons such as fish physiological and endocrine status, sexual maturation and spawning, water temperature and other biotic, abiotic and anthropogenic factors.

2.8 Other Biomarkers

The exposure of the organism to xenobiotics provokes the genetic alterations which initiate structural changes of the DNA, expression of the mutant genes and carcinogenesis. The integrity of DNA can be greatly affected by genotoxic chemicals due to DNA strand breaks, loss of methylation and formation of DNA adducts. DNA strand break may occur due to direct DNA damage by exogenous xenobiotics or may be produced during DNA repair process or other physiological responses in the cell. Organic chemicals interact with DNA to form both stable and unstable adducts with DNA and each type of adduct may contribute towards the eventual transformation of the cell (van der Oost et al., 2003; Sarkar et al., 2006). There is strong correlation between the level of chemical binding to DNA and carcinogenic potency and organ specificity. Apart from pollutant levels, these factors may be related to the age and sex of the organisms, as well as season, water temperature and food availability. Both laboratory and field studies on DNA adduct formation have been reviewed by van der Oost et al. (2003). The DNA adduct responses for all fish species from 17 laboratory studies and 30 field studies showed a significant increase in hepatic DNA adduct levels in 100% of the laboratory studies and 70% of the field studies, while strong increases (> 500% of the control) were observed in 65% and 30% of the laboratory and field studies, respectively. Thus, DNA damage analysis is a good biomarker for the evaluation of fish polluted effects.

The immune system comprises families of cells capable for rapid proliferation and differentiation, regulated by a variety of soluble factors. Immunity is very sensitive to unfavorable impact and may be used as potential biomarker in fish biomonitoring (van der Oost et al., 2003). Polycyclic aromatic hydrocarbons (PAH) are an important class of environmental pollutants that are known to be carcinogenic and immunotoxic. Among innate immune parameters, many authors have focused on macrophage activities in fish exposed to PAH. Macrophage respiratory burst appears especially sensitive to PAH. Among acquired immune parameters, lymphocyte proliferation appears highly sensitive to polycyclic aromatic hydrocarbon exposure. However, the effects of PAH on both specific and non-specific immunity are contradictory and depend on the mode of exposure, the dose used or the species tested. In contrast to mammals, fewer studies have been done in fish to determine the mechanism of PAH-induced toxicity. This phenomenon seems to implicate different intracellular mechanisms such as metabolism by cytochrome P4501A, binding to the Ah-receptor, or increased intracellular calcium. Advances in basic knowledge of fish immunity should lead to improvements in monitoring fish health and predicting the impact of polycyclic aromatic hydrocarbons on fish populations, which is a fundamental ecotoxicological goal (Reynaud and Deschaux, 2006).

Oxidative molecules including lipids, amino acids and proteins could be used as potential biomarkers of negative biological effects on fish in polluted locations. The use of protein CO groups as biomarkers of oxidative stress-led pollution has some advantages in comparison to the measurements of other oxidation products because of the relative early formation and the relative stability of carbonilated and oxidized proteins (Dalle-Donne et al., 2003). We noted the correlation between the concentration of Pb and Hg in fish tissues and the level of oxidized proteins in blood serum. In addition, the level of the oxidized proteins and olygopeptides was significantly higher in fish from polluted locations as compared to non-polluted habitats (Rudneva et al., 2011).

There are some other biomarkers which may successfully be used in fish monitoring and it is very important to continue to search another ones for understanding the polluted effects in fish and the mechanisms of their resistance to unfavorable factors of the environment. Physiological and biochemical assays are useful for indicating the "early warning" biomarkers which induce at the beginning of stress reaction. Such biomarkers affect in very low stressors concentrations and their fluctuations should be monitored and compared with the normal ranges. More research is required for search, application and evaluation of significance of stress biomarkers in fish impacted polluted environments.

References

Abd-Algadir, M.I., M.K.S. Elkhier and O.F. Idris. 2011. Changes of fish liver (*Tilapia nilotica*) made by herbicide (*Pendimethalin*). J. Applied Biosciences. 43: 2942–2946.

Abdel-Hameid, N.-A.H. 2007. Physiological and histopathological alterations induced by phenol exposure in *Oreochromis aureus* juveniles. Turkish J. Fisheries. Aquatic Sciences. 7: 131–138.

Adams, S.M. 2005. Assessing cause and effect of multiple stressors on marine systems Mar. Pollut. Bull. 51 (8–12): 649–657.

Adedeji, O.B. 2011. Response of acetylcholinesterase activity in the braib of *Clarias gariepinus* to sublethal concentration of diazinon. J. Appl. Sci. Environ. Sanitation. 6(2): 137–141.

Ahmad, I., M. Pacheco and M.A. Santos. 2004. Enzymatic and nonenzymatic antioxidants as an adaptation to phagocyte-induced damage in *Anguilla anguilla* L. following *in situ* harbor water exposure. Ecotoxicol. and Environ. Safety. 57: 290–302.

Ajani, F., B.O. Emikpe and O.K. Adeyemo. 2011. Histopathological and enzyme changes in Clarias gariepinus (Burchell, 1822) exposed in nitrite at different water temperatures. Nature and Science. 9(5): 119–124.

Amado, L.L., C.E. da Rosa, A.M. Leite, L. Moraes, W.V. Pires, G.L.L. Pinho, C.M.G. Martins, R.B. Robaldo, L.E.M. Nery, J.M. Monserrat, A. Bianchini, P.E. Martinez and L.A. Geracitano. 2006a. Biomarkers in croakers *Micropogonias furnieri* (Teleostei: Sciaenidae) from polluted and non-polluted areas from Patos Lagoon estuart (Southern Brazil): evidences of genotoxic and immunological effects. Mar. Pollut. Bull. 52: 199–206.

Amado, L.L., R.B. Robaldo, L.A. Geracitano, J.M. Monserrat and A. Bianchini 2006b. Biomarkers of exposure and effect in the Brazilian flounder *Paralichthys orbignyanus* (Teleostei: Paralichthyidae) from the Patos Lagoon estuary (Southern Brazil): Evidences of genotoxic and immunological effects. Mar. Pollut. Bull. 52: 207–213.

Arukwe, A. and A. Goksoyr. 2003. Eggshell and egg yolk proteins in fish: hepatic proteins for the next generation: oogenetic, population, and evolutionary implications of endocrine disruption. Comp. Hepatology. 2(4): 1–21.

Assis, C.R.D., R.S. Bezerra and L.B. Carvalho. 2011. Fish cholunesterases as biomarkers of organophosphorus and carbamate pesticides. In: Pesticides in the Modern World–Pests Control and pesticides Exposure and Toxicity Assessment. (Ed. M. Stoytcheva). In Tech. Publ. 2011. Chapter 13, pp. 254–278

Bainy, A.C.D., E. Saito, S.M. Carvalho and V.B.C. Junquerira. 1996. Oxidative stress in gill, erythrocytes, liver and kidney of Nile tilapia (*Oreochromis niloticus*) from a polluted site. Aquatic. Toxicol. 34: 151–162.

Baker, M.E. 2002. Albumin, steroid hormones and the origin of vertebrates. Journal of Endocrinology. 175: 121–127.

Baker, S.M. and R. Mann. 1991. Metabolic rates of metamorphosis oysters (*Crassostrea virginica*) determined by microcalorimetry. Am. Zool. 31: 134 A.

Beyers, D.W., J.A. Rice, W.H. Clements and C.J. Henry. 1999. Estimating physiological coast of chemical exposure: integrating energetics and stress to quantify toxic effects in fish. Can J. Fish. Aquat. Sci. 56: 814–822.

Bierkens, G.E.A. 2000. Application and pitfalls of stress-proteins in biomonitoring. Toxicology. 153(1-3): 61–72.

Boone, J.S. and J.E. Chambers. 1995. Time course of inhibition of cholinesterase and aliesterase activities, and bobprotein sulfhydryl levels following exposure to organophosphorus insecticedes in mosquitofish (*Gambusia affinis*). Fundamental Applied Toxicol. 29: 202–207.

Bretaud, S., P. Saglio and J.P. Toutant. 2001. Effects du carbofuran sur l'activite de l'acetylcholinesterase cerebrale et sur l'activite de nage *Carassius auratus* (Cyprinidae). Cybium. 25: 33–40.

Burlakova, E.B. 2005. Bioantioxidants: yesterday, today, tomorrow. In: Chemical and biological kynetics. New approach. Moscow. Chemistry Publ. 2: 10–45 (*in Russian*).

Canepa, E., S. Fraschetti, S. Geraci, M. Licciano, M. Manganelli, G. Alberyelli and G. Riadi. 1997. Microcalorimetry of some invertebrates: preliminary characterization of their metabolic activity during different developmental stages. Biol. Mar. Mediterr. 4: 626–628.

Canesi, L., A.Viarengo, C. Leonzio, M. Filippelli and G. Galllo. 1999. Heavy metals and glutathione metabolism in mussel tissues. Aquat. Toxicol. 46: 67–76.

Canty, M.N., J.A. Hagger, R.T.B. Moore, L. Cooper and T.S. Galloway. 2007. Sublethal impact of short term exposure to the organophosphate pesticide azamethiphos in marine mollusk *Mytilus edulis*. Marine Pollution Bull. 54: 396–402.

Coppo, J.A., N.B. Mussart and S.A. Fioranelli. 2001–2002. Physiological variation of enzymatic activities in blood of bullfrog, *Rana catesbeianna* (Shaw, 1802). Rev. Vet. 12/13(1-2): 22–27.

Currie, S., B.L. Tufts and C.D. Moyes. 1999. Influence of bioenergetic stress on heat shock protein gene expression in nucleated red blood cells in fish. Am. J. Physiol. Regul. Integr. Comp. Physiol. 276: R990–R996.

Dafre, A.L. and E. Reischl. 1990. High hemoglobin mixed disulfide content in hemolysates from stressed shark. Comp. Biochem. Physiol. 96B: 215–219.

Dai, H.Q., F.W. Edens and R.W. Roe. 1996. Glutathione-S-transferases in the Japanese quail: tissue distribution and purification of the liver isoenzymes. J. Biochem. Mol. Toxicol. 11: 85–96.

Dalle-Donne, I., R. Rosi, D. Giostarini, A. Milzani and R. Colombo. 2003. Protein carbonyl groups as biomarkers of oxidative stress. Clinca. Chimica Acta. 329: 23–38.

De La Fuente, M., A. Hermanz and M.C. Vallejio. 2005. The immune system in the oxidative stress conditions of aging and hypertension: favorable effects of antioxidants and physical exercise. Antioxidants & Redox Signaling. 7: 1356–1366.

De Smet, H., R. Blust and L. Moens. 1998. Absence of albumin in the plasma of the common carp *Cyprinus carpio*: binding of fatty acids to high density lipoprotein. Fish Physiol. and Biochem. 19: 71–81.

Dhainoaut, A., J. Bonaly, J.-Ph. Baeque, C. Minier and Th. Caquet. 1997. Proteines de shock thermique et resistance multixenobiotique. In Biomarqeures en ecotoxicologie. Ed. Lagadic L., Caquet Th., Amiard J-C. and Ranade F. Masson, Paris. 67–95.

Di Giulio, R.T., P.C. Washburn, R.J. Wenning, G.W. Winston and C.S. Jewell. 1989. Biochemical responses in aquatic animals: a review of determination of oxidative stress. Environ. Toxicol. and Chem. 8: 1103–1123.

Egaas, E., J.G. Falls, N.O. Svendsen, H. Ramstad, J.U. Shkaare and W.C. Dauterman. 1995. Strain- and sex-specific differences in the glutathione S-transferase class pi in the mouse examined by gradient elution of the glutathione-affinity matrix and reverse-phase high performance liquid chromatography. Biochimica et Biophysica Acta. 1243: 256–264.

Eidem, J.K., H. Kleivdal, K. Kroll, N. Denslow, R. van Aerle, Ch. Tyler, G. Panter, T. Hutchinson and A. Goksoyr. 2006. Development and validation of a direct homologous quantitative sandwich ELISA for fathead minnow (*Pimephales promelas*) vitellogenin. Aquatic Toxicol. 78: 202–206.

El-Seify, M.A., M.S. Zaki, A.R.Y. Desouky, H.A. Abbas, O.K.A. Hady and A.A.A. Zaid. 2011. Study on clinopathological and biochemical changes in some freshwater fishes infected with external parasites and subjected to heavy metals pollution in Egypt. Life Science J. 8(3): 401–408.

Farombi, E.O., O.A. Adelowo and Y.R. Ajimoko. 2007. Biomarkers of oxidative stress and heavy metal levels as indicators of environmental pollution in African cat fish (*Clarias gariepinus*) from Nigeria Ogun River. Environmental Research and Public Health. 4(2): 158–165.

Filho, W.D. 2007. Reactive oxygen species, antioxidants and fish mitochondria. Fishery in Bioscience. 12: 1229–1237.

Filho, D.W. and A. Boveris. 1993. Antioxidant defences in marine fish—II. Elasmobranchs. Comp. Biochem. Physiol. 106C: 415–418.

Filho, D.W., C. Giulivi and A. Boveris. 1993. Antioxidant defences in marine fish—I. Teleosts. Comp. Biochem. Physiol. 106C: 409–413.

Filho, D.W., M.A. Torres, T.B. Tribess, R.C. Pedrosa and C.H.L. Soares. 2001. Influence of season and pollution on the antioxidant defences of the cichlid fish acara (*Geophagus brasiliensis*). Brasilian J. Medical and Biological Res. 34: 719–726.

Filho, W.D., M.A. Torres, E. Zaniboni-Filho and R.C. Pedrosa. 2005. Effect of different oxygen tensions on weight gain, feed conversion and antioxidant status in piapara *Leporinus elongates* (Valenciennes, 1847). Aquaculture. 244: 349–357.

Fossi, M.C., S. Casini, L. Marsili, G. Neri, G. Mori, S. Ancora, A. Mascateli, A. Ausili and G. Notarbartolo-di-Sciara. 2002. Biomarkers for endocrine disruptors in three species of Mediterranean large pelagic fish. Mar. Environ. Res. 54: 667–671.

Fulton, M.H. and P.B. Key. 2001. Acetylcholinesterase inhibition in estuarine fish and invertebrates as an indicator of organophosphorus insecticide exposure and effects). Environ. Toxicol. Chem. 20(1): 37–45.

Galloway, T.S. 2006. Biomarkers in environmental and human health risk assessment. Mar. Pollut. Bull. 53: 606–613.

George, S.G. 1994. Enzymology and molecular biology of phase II xenobiotic-conjugating enzymes in fish. In: D.S. Malins and G.K. Ostrander (eds.). Aquatic Toxicology; Molecular, Biochemical and Cellular perspectives. Lewis Publishers, CRC press, pp. 37–85.

Georgiades, E.T. and D.A. Holdway. 2007. Optimization of acetylcholinestarase and metabolic enzyme activity in multiple fish species. Water Qual. Res. J Canada. 42(2): 101–100.

Goksoyr, A. 2006. Endocrine disruptors in the marine environment: mechanisms of toxicity and their influence on reproductive processes in fish. J. Toxicol. Environ. Health. Part A. 69(1-2): 175–184.

Goksoyr, A., J. Beyer, E. Egaas, B.E. Grosvik, K. Hylland, M. Sandvik and J.U. Skaare. 1996. Biomarker responses in flounder (*Platichthys flesus*) and their use in pollution monitoring. Mar. Pollut. Bull. 33: 36–45.

Goto, T., H. Sugiyama, H. Funatsu, Y. Osada, Sh. Takagi and A. Mochzuki. 2003. Measurement and distribution of hepatic cysteinesulfinate aminotransferase activity in fish. Susanzoshoku. 51(3): 361–362.

Grune, T. Oxidative stress, aging and the proteasomal system. 2000. Biogerontology. 1: 31–40.

Gwozdzinski, K., H. Roche and G. Peres. 1992. The comparison of effects of heavy metal ions on the antioxidant enzyme activities in human and fish *Dicentrarchus labrax* erythrocytes. Comp. Biochem. Physiol. 102C: 57–60.

Habig, W.H., M.J. Pabst and W.B. Jokoby. 1974. Glutathione S-transferase. The first enzyme step in mercapturic acid formation. J. Biol. Chem. 249: 7130–7139.

Haegan, W.M., H. Vander, R.B. Owen and W.B. Krohn. 1994. Metabolic rate of American woodcock. Wilson Bulletin. 106(2): 338–443.

Hamed, R.R., N.M. Farid, Sh. E. Elowa and A.-M. Abdalla. 2003. Glutathione related enzyme levels of freshwater fish as bioindicators of pollution. The Environmentalist. 23: 313–322.

Hamed, R.R., M. Maharem and R.A.M. Guinidi. 2004. Glutathione and its related enzymes in the Nile Fish. Fish Physiology and Biochemistry. 30: 189–199.

Handy, R.D. and M.H. Depledge. 1999. Physiological responses: their measurements and use as environmental biomarkers in ecotoxicology. Ecotoxicology. 8: 329–349.

Hansson, T., D. Schiedek, K.K. Lehtonen, P.J. Vuorinen, N.E. Liewenborg, U. Tjarnlund, M. Hansson and L. Balk. 2006. Biochemical biomarkers in adult female perch (*Perca fluviatilis*) in a chronically polluted gradient in the Stockholm recipient (Sweden). Marine Pollution Bulletin. 53: 451–468.

Hardy, R.W. 2001. Nutritional deficiency in commercial aquaculture: likelihood, onset and identification. In: Nutrition and Fish Health (Ed. Ch.Lim and Webster C.D.). N.Y., Oxford. 131–148.

Hart, H.U. 1996. Molecular chaperons in cellular protein folding. Nature. 381: 571–579.

Heikkila, J.J., G.A. Schultz, K. Iatrou and L. Gedamu. 1982. Expression of a set of fish genes following heat or metal ion exposure. J. Biol. Chem. 25: 12000–12005.

Hiramatsu, N., T. Matsubara, T. Fujita, C.V. Sullivan and A. Hara. 2006. Multiple piscine vitellogenins: biomarkers of fish exposure to estrogenic endocrine disruptors in aquatic environments. Marine Biol. 149(1): 35–47.

Hopkins, W.A., C.P. Tatara, H.A. Brant and C.H. Jagoe. 2003. Relationships between mercury body concentrations, standard metabolic rate and body mass in castem mosquitofish *Gambusia holbrooki* from three experimental pollution. Environ. Toxicol. Chem. 22: 586–590.

Hu, D., E. Klann and E. Thiels. 2007. Superoxide dismutase and hippocampal function: age and isozyme matter. Antioxidants & Redox Signaling. 9: 201–210.

Hussein, G., U. Sankawa, H. Goto, K. Matsumoto and H. Watanabe. 2006. Astaxantin, a carotenoid with potential human health and nutrition. J. Nat. Prod. 69: 443–449.

Hutchinson, T.H., R. Brown, K.E. Brugger, P.M. Campbell, M. Holt, R. Lange, P. McCahon, I.J. Tattersfield and R. van Egmond. 2000. Ecological risk assessment of endocrine disruptors. Environ. Health Perspectives. 108(11): 1007–1014.

Hutchinson, T.H., G.T. Ankly, H. Segner and R.T. Tyler. 2006. Screening and testing for endocrine disruption in fish—biomarkers as "signposts", not "traffic lights" in risk assessment. Environ. Health Perspectives. 114(1): 106–114.

Ishioka, H. and T. Fushimi. 1975. Some haematological properties of matured Red bream, *Chrysophrys major*. Bull. Nansei. Reg. Fish. Res. Lab. 8: 11–20.

Iwama, G.K., M.M. Vuayan, R.B. Forsyth and P.A. Ackerman. 1999. Heat shock proteins and physiological stress in fish. Amer. Zool. 39: 901–909.

Iwama, G.K., L.O.B. Afonso, A. Todgham, P. Ackerman and K. Nakano. 2004. Are hsps suitable for indicating stressed states in fish? J. Experim. Biol. 207: 15–19.

Keckeis, H. and F. Schimer. 1 990. Consumption, growth and respiration of bleak, *Alburnus alburnus* (L.), and roach *Rutilus rutilus* (L.) during early ontogeny. J. Fish Biol: 36: 841–851.

Kirby, M.F., S. Morris, M. Hurst, S.J. Kirby, P. Neall, T. Tylor and A. Fagg. 2000. The use of holinesterase activity in flounder (*Platichthys flesus*) muscle tissue as a biomarker of neurotoxic contamination in UK estuaries. Mar. Pollut. Bull. 40(9): 780–791.

Klemz, C., L.M. Salvo, J.C.B. Neto, A.C.D. Bainy and C.S. Assis. 2010. Cytochrome P450 detection in liver of the catfish *Ancistrus multispinis* (Osteichthyes, Loricariidae), Brazilian Archives of Biology and Technology. 53(2): 361–368.

Lali, S.P. and L.M. Lewis-McCrea. 2007. Role of nutrients in skeletal metabolism and pathology in fish—an overview. Aquaculture. 267: 3–19.

Lamprecht, I. 1998. Monitoring metabolic activities of small animals by means of microcalorimetry. Pure & Appl. Chem. 70(3): 695–700.

Langenbuch, M. and H.O. Portner. 2003. Energy budget of hepatocytes from Antactic fish (*Pachycara brachycephalum* and *Lepidonototnen kempi*) as a function of ambient CO_2: pH-dependent limitation of cellular protein biosynthesis? J. Experim. Biol. 206: 3895–3903.

Leaver, J., K. Scott and S.G. George. 1993. Cloning and characterization of the major hepatic gluthathione-S-transferase from a marine teleost flatfish, the plaice (*Pleuronectes platessa*) with structural similarities to plant, insect and mammalian Theta class isoenzymes. Biochem. J. 292: 189–195.

Lesser, M.P. 2006. Oxidative stress in marine environments: biochemistry and physiological ecology. Ann. Rev. Physiol. 68. 253–278.

Li, S.-D., Y.-D. Su, M. Li and Ch.-G. Zou. 2006. Hemin-mediated hemolisis in erythrocytes: effects of ascorbic acid and glutathione. Acta Biochem. Biophys. Sinica. 38(1): 63–69.

Livingstone, D.R. 2001. Contaminant-stimulated reactive oxygen species production and oxidative damage in aquatic organisms. Mar. Pollut. Bull. 42: 656–665.

Lun, L.-M., Sh.-C Zhang and Y.-J. Liang. 2006. Alanine aminotransferase ion amphioxus: presence, localization and Up-regulation after acute lipopolysaccharide exposure. J. Biochem. Molecular Biol. 39(5): 511–515.

Lyndon, A.R., D.F. Houlihan and S.J. Hall. 1992. The effect of short term fasting and single meal on protein synthesis and oxygen consumption in cod, *Gadus morphua*. J. Comp. Physiol. Biochem. 162B: 209–215.

Maenpaa, K.A., Penttinen O.-P. and J.V.K. Kukkonen. 2004. Pentachlorophenol (PCP) bioaccumulation and effect of heat production on salmon eggs at different stages of development. Aquatic Toxicology. V. 68(1): P75–85.

Maenpaa, K.A., M.T. Leppanen and J.V.K. Kukkonen. 2009. Sublethal toxicity and biotransformation of pyren in *Lumbriculus variegates* (Oligochaeta). Science of the Total Environment. 407: 2666–2672.

Magnotti, R.A., J.P. Zaino and R.S. MaConnell. 1994. Pesticide-sensitive fish muscle cholinesterases. Comp. Biochem. Physiol. 108C(2): 187–194.

Martinez-Alvarez, R., A. Morales and A. Sanz. 2005. Antioxidant defenses in fish: biotic and abiotic factors. Rev. Fish Biology & Fisheries. 15: 75–88.

Martinez-Gomez, C., B. Fernandez, J. Valdes, J. A. Campillo, J. Benendicto, F. Sanchez and A.D. Vethaak. 2009. Evaluation of three-year monitoring with biomarkers the *Prestige* oil spill (Spain). Chemosphere. 74(5): 613–620.

Martinez-Porchas, M., M. Hernandez-Rodrigues., J. Davila-Ortiz, V. Vila-Cruz and J.R. Ramos-Enriquez. 2011. A preliminary study about the effect of benzo[a]pyrene (BaP) injection on the thermal behavior and plasmatic parameters of the Nile tilapia ((*Oreochromis niloticus* L.) acclimated to different temperatures. Panamjas (Pan-American J. Aquatic Sciences). 6(1): 76–85.

Mc Donald, D.G. and C.L. Milligan. 1992. Chemical properties of the blood. In: Fish Physiology. V. XII. Part B.: The cardio vascular system. Ed. By W.S. Hour, D.J. Randall and Farrell A.P. Academic Press N.Y. 55–135.

McKenzie, D.J., E. Garofalo, M.J. Winter, S. Ceradini, P. Verweij, N. Day, R. Hayes, R. van der Oost, P.J. Butler, J.K. Chipman and E.W. Taylor. 2007. Complex physiological traits as biomarkers of the sub-lethal toxicological effects of pollutant exposure in fishes. Phil. Trans. R. Soc. 362B: 2043–2059.

Mehdi, Y., Sh. Amiri and M. Kor Davood. 2011. Serum biochemical parameter of male, immature and female Persian Stugeon (*Acipencer persian*). Australian J. of Basic and Applied Sci. 5(5): 436–481.

Metcalf, V.J. and N.J. Gemmell. 2005. Fatty acid transport in cartilaginous fish: absence of albumin and possible utilization of lipoproteins. Fish Physiol. and Biochem. 31: 55–64.

Moeyner, K. 1993. Changes in serum protein composition occur in Atlantic salmon, *Salmo salar* L. during *Aeromonas salmonicidae* infection. J. Fish Diseases. 16(6): 601–604.

Morcillo, Y., G. Janer, S.C.M. O'Hara, D.R. Livingstone and C. Porte. 2004. Interaction of tributylin with hepatic cytochromeP450 and uridine diphosphate-glucuronosyl transferase systems of fish: *in vitro* studies. Environ. Toxicol. Chem. 23(4): 990–996.

Moseley, P.L. 1997. Heat shock proteins and heat adaptation of the whole organism. J. Appl. Physiology. 83(5): 1413–1417.

Moussa, F.I., M.B. Abou-Shabana and M.Y. El-Toweissay. 1994. Short and long term, effects of acid stress on survival, behaviour, and some cellular blood constituents in the cat fish *Clarias lazera*. Bull. Natl. Inst. Oceanogr. Fish. Egypt. 20(2): 229–218.

Napierska, D., J. Barsiene, E. Mulkiwicz, M. Podolska and A. Rybakovas. 2009. Biomarker responses in flounder *Platichthys flesus* from the Polish coastal area of the Baltic Sea and applications in biomonitoring. Ecotoxicology. 18: 846–859.

Nicolas, J.-M. 1999. Vitellogenesis in fish and the effects of polycyclic aromatic hydrocarbon contaminants. Aquatic Toxicol. 45(2-3): 77–90.

Nimmo, I.A. 1987. The glutathione S-transferase of fish. Fish Physiol. Biochem. 3(4): 163–172.

Normat, M., G. Graf and A. Szaniawska. 1998. Heat production in Saduria entomon (Isopoda) from Gulf of Gdansk during an experiment exposed to anoxia conditions. Marine Biol. 131: 269–273.

Oost van der, R., J. Beyer and N.P.E. Vermeulen. 2003. Fish bioaccumulation and biomarkers in environmental risk assessment: a review. Environmental Toxicology and Pharmacology. 13: 57–149.

Otto, D.M.E. and T.W. Moon. 1996. Endogenous antioxidant systems of two teleost fish, the rainbow trout and the black bullhead, and the effect of age. Fish Physiol. Biochem. 15: 349–358.

Oven, L.S., I.I. Rudneva and N.F. Shevchenko. 2000. Biomarkers application in the evaluation of the BlackSea *Spicara flexuosa* (Centracanthidae) health. J. Ichthyology. 40: 810–813 (*in Russian*).

Oven, L.S., I.I. Rudneva and N.F. Shevchenko. 2001. The present state of the Mediterranean atherina *Atherina hepsetus* population in the coastal area of the Black Sea. J. Ichthyology. 42: 425–428 (*in Russian*)

Ozmen, I., A. Bayir, M. Cengiz, A.N. Sirkecioglu and M. Atamanalp. 2004. Effects of water reuse system on antioxidant enzymes of rainbow trout (*Oncorhynchus myksiss* W., 1792). Vet. Med. Czech. 49: 373–378.

Ozmen, I., M. Atamanalp, A. Bayir, A.N. Sirkecioglu and M. Cengiz. 2007. The effects of different stressors on antioxidant enzyme activities in the erythrocyte of rainbow trout (*Oncorhynchus myksiss* W., 1792). Fresenius Environ. Bull. 16: 922–927.

Padmini, E. 2012. Heat shock proteins modulate signaling pathways in survival of stressed fish to polluted environments. In: Fish Ecology. (Ed. Gempsy P.) Nova Science Publishers. New York, pp. 173–193.

Pane, E.F., C. Smith, J.C. McGreer and C.M. Wood. 2003. Mechanism of acute and chronic waterborne nickel toxicity in the freshwater cladoceran Daphnia magna. Environ. Dci. Technol. 37: 4382–4389.

Parra, G. and M. Yufera. 2001. Comparative energetics during early development of two marine fish species *Solea senegalensis* (Kaup) and *Sparus aurata* (L.). J. Exp. Biol. 204: 2175–2183.

Penttinen, O.P. and J.V.K. Kukkonen. 1998. Chemical stress and metabolic rate in aquatic invertebrates: threshold, dose-response and mode of toxic action. Environ. Toxicol. Chem. 17. 883–890.

Penttinen, O.P., J.O. Honkanen, K. Sorsa and J.V.K. Kukkonen. 2005. Can aquatic pollutants cause specific endocrinological and metabolic responses in salmon (*Salmo salar* m. Sebago) embryos? A direct calorimetry study. Verhandlungen International Vereinigung Limnology. 29: 945–948.

Penttinen, O.-P. and J.V.K. Kukkonen. 2006. Body residues as dose for sublethal responses in alevins of landlocked salmon (*Salmo salar* m Sebago): a direct calorimetry study. Environ. Toxicol. Chem. 25(4): 1088–1093.

Penttinen, O.-P. and J. Holopainen. 1995. Physiological energetics of a midge, *Chironomus riparius* Meigen (Insecta, Diptera): normoxic heat output over the whole life cycle and response of larva to hypoxia and anoxia. Oecologia. 103: 419–424.

Penttinen, O.P., J. Kukkonen and J. Pellinen. 1996. Preliminary study to compare residues and sublethal energetic responses in benthic invertebrates exposed to sediment-bound 2,4,5-trichlorophenol. Environ. Toxicol. Chem. 15: 160–166.

Peters, L.D. and D.R. Livingstone. 1996. Antioxidant enzyme activities in embryonic and early larval development of turbo. J. Fish Biol. 49: 986–997.

Porte, S., G. Janer, L.C. Lorusso, M. Ortiz-Zarragoitia, M.P. Cajaraville, M.C. Fossi and L. Canesi. 2006. Endocrine disruptors in marine organisms: approaches and perspectives. Comp. Biochem. Physiol. 143C: 303–315.

Rajamanickam, V. and N. Muthuswamy. 2008. Effect of heavy metals induced toxicity on metabolic biomarkers in common carp (*Cyprinus carpio* L). Mj. Int. J. Sci. Tech. 2(10): 192–200.

Razo, del L.M., B. Quintanilla-Vega, E. Brambila-Colombres, E.S. Calderon-Arand, M. Manno and A. Albores. 2001. Stress proteins induced by arsenic. Toxicol. Applied Pharmacol. 177(2): 132–148.

Reynaud, S. and P. Deschaux. 2006. The effects of polycyclic aromatic hydrocarbons on the immune system of fish: A review. Aquatic Toxicology. 77(1-2): 229–238.

Richards, J.G. 2010. Metabolic rate suppression as a mechanism for surviving Environmental challenge in fish. In: Progress in Molecular and Subcellular Biology. 49: 113–135.

Richmonds C.R. 1990. Effects of malathion on some physiological, histological and behavioral aspects of bluegill sunfish *Lepomis macrochirus*. Diss. PhD. DA9014838 Kent State University. 51(1): 161 pp.

Roshina, O.V. and I.I. Rudneva. 2009. Evaluation of fungicide cuprocsat toxicity used fish biomarkers. Toxicol. Rev. 4: 26–31 (*in Russian*).

Roche, H. and G. Boge. 1996. Resposes des activites antioxydantes erythrocytaires du loup (*Dicentrarchus labrax*) a des modifications experimentales de parameters chimiques ou physiques du milieu. Mar. Life. 6: 53–61.

Rocha-e-Silva, T.A.A., M.M. Rossa, F.T. Rantin, T. Matsumura-Tundisi, J.G. Tundisi and I.A. Degterev. 2004. Comparison of liver mixed-function oxygenase and antioxidant enzymes in vertebrates. Comp. Biochem. Physiol. 137C: 155–165.

Rudneva, I.I. 1997. Blood antioxidant system of Black Sea elasmobranch and teleosts. Comp. Biochem. Physiol. 118C: 255–260.

Rudneva, I.I. 1998. Ecological and phylogenetic peculiarities of lipid composition and lipid peroxidation in teleosts and elasmobranch in Black Sea. J. Evol. Biochem. Physiol. 34: 310–318 (*Saint-Petersburg, in Russian*).

Rudneva, I.I. 2011. Ecotoxicological Studies of the Black Sea Ecosystem. The Case of Sevastopol Region. Nova Science Publishers, Inc. New York, USA. 2011. 62 pp.

Rudneva, I.I. 2012. Antioxidant defence in marine fish and its relationship to their ecological status. In: Fish Ecology. (ed. Dempsy S.P.) Nova Science Publishers, Inc. New York, USA. P. 31–59.

Rudneva, I.I., T.L. Chesalina, V.G. Shaida and N.F. Shevchenko. 1998. Morphology and heat production in atherina larvae (*Atherina hepsetus* L.) from contaminated and non-contaminated regions. Marine Ecology. 47: 33–36 (*in Russian*).

Rudneva, I.I. and N.V. Zherko. 1999. The effect of PCB on antioxidant system and lipid peroxidation in gonads of Black Sea red mullet *Mullus barbatus ponticus*. Marine Biology. 25: 239–242 (*in Russian*).

Rudneva, I.I. and N.V. Zherko. 2000. The effect of PCB on antioxidant system and lipid peroxidation in gonads of Black Sea Scorpion fish. Ecology. 1: 70–73 (*in Russian*).

Rudneva, I.I. and E. Petzold-Bradley. 2001. Environmental and security challenges in the Black Sea region. In: E. Petzold-Bradley, A. Carius and A. Vimce (eds.). Environment Conflicts: Implications for Theory and Practice. Netherlands: Kluwer Academic Publishers. pp. 189–202.

Rudneva, I.I., N.F. Shevchenko, I.N. Zalevskaya and N.V. Jerko. 2005. Biomonitoring of the oastal waters of Black Sea. Water Resources. 32(2): 238–246 *(in Russian)*.

Rudneva, I.I., E.N. Skuratovskaya, S.O. Omelchenko and I.N. Zalevskaya. 2008a. The application of fish blood biomarkers for the ecotoxicological evaluation of marine coastal waters. Ecological Chemistry. 17(2): 77–84 *(in Russian)*.

Rudneva, I.I., O.V. Roshina, S.O. Omelchenko and I.N. Zalevskaya. 2008b. Use fish bioindicators for the analysis of seasonal dynamics of marine environment ecological status. Ecological Chemistry. 17(3): 24–29 *(in Russian)*.

Rudneva, I.I., N.S. Kuzminova, E.N. Skuratovskaya and T.B. Kovyrshina. 2010a. Comparative Study of Glutathione-S-transferase Activity in Tissues of Some Black Sea Teleosts. International J. of Science and Nature. 1(1): 1–6.

Rudneva, I.I., N.S. Kuzminova and E.N. Skuratovskaya. 2010b. Glutathion-S-transferases activity in tissues of Black Sea fish species. Asian J. Exp. Biol. Sci. 1(1): 141–150.

Rudneva, I.I., E.N. Skuratovskaya, N.S. Kuzminova and T.B. Kovyrshina. 2010c. Age composition and antioxidant enzyme activities in blood of Black Sea teleosts. Comp. Biochem. Physiol. 151C: 229–239.

Rudneva, I.I. and N.S. Kuzminova. 2011. Effect of Chronic Pollution on Hepatic Antioxidant System of Black Sea Fish Species. Int. J. Science and Nature. 2(2): 279–286.

Rudneva, I.I., E.N. Skuratovskaya, I.I. Dorohova, J.A. Grab, I.N. Zalevskaya and S.O. Omelchenko. 2011. Bioindication of the environmental state of marine water areas with the use in fish biomarkers. Water Resources. 38(1): 104–109.

Rudneva, I.I. and T.B. Kovyrshina. 2011. Comparative study of electrophoretic characteristics of serum albumin of round goby *Neogobius melanostomus* from Black Sea and Azov Sea. International Journal of Advanced Biological Research. 1(1): 131–136.

Rudneva, I.I. and V.G. Shaida. 2012. Metabolic rate of marine fish in early life and its relationship to their ecological status. In: Fish Ecology (Ed. Gempsy P.). Nova Science Publishers. New York. pp. 1–29.

Rudneva, I.I., E.N. Skuratovskaya, I.I. Dorohova and T.B. Kovyrshina. 2012. Use of fish blood biomarkers for evaluation of marine environment health. World J. Science and Technology. 2(7): 19–25.

Russel, M., J. Yao, H. Chen, F. Wang, Y. Zhou, M.M. Choi, G. Zaray and P. Trebse. 2009. Different technique of microcalorimetry and their application to environmental sciences: a review. J. American Science. 5(4): 194–208.

Sanders, B.M. 1993. Stress proteins in aquatic organisms: an environmental perspectives.// Crit. Rev. in Toxicol. 23(1): 49–75.

Sanders, B.M., L.S. Martin, W.G. Nelson, D.K. Pheleos and W. Welch. 1991. Relationship between accumulation of a 60 kDa stress protein and scope-for-growth *Mytilus edulis* exposed to a range of cooper concentrations Mar. Environ. Res. 31: 91–97.

Sanders, B.M., L.S. Martin, S.R. Howe, W.G. Nelson, E.S. Hegre and D.K. Phelps. 1994. Tissue specific differences in accumulation of stress proteins in *Mytilus edulis* exposed to a range of cooper concentrations. Toxicol. Appl. Pharmacol. 125. 206–213.

Sarkar, A., D. Ray, A.N. Shrivastava and S. Sarker. 2006. Molecular biomarkers: their significance and application in marine pollution monitoring. Ecotoxicology. 15: 333–340.

Sayed, A. El-Din H., I.A.A. Mekkawy and U.M. Mahmoud. 2011. Effects of 4-nonylphenol on metabolic enzymes, some ions and biochemical blood parameters of the African catfish *Clarias gariepinus* (Burchell, 1822). Afr. Biochem. Res. 5(9): 287–297.

Schlotfeldt, H.J. 1975. Evidence of seasonal variations of serum proteins of rainbow trout (Salmo gairdneri Rich) by cellulose acetate electrophoresis. Zentralbl Veterinaermedb. 22(2): 113–129.

Serensen, J.G., T.N. Kristensen and V. Loeschcke. 2003. The evolutionary and ecological role of heat shock proteins. Ecology Lett. 6. 1025–1037.

Sole, M., S. Rodriguez, V. Papiol, F. Maynou and J.E. Cartes. 2009. Xenobiotic metabolism markers in marine fish with different trophic strategies and their relationship to ecological variables. Comp. Biochem. Physiol. 149C(1): 83–89.

Soti, C. and P. Csermely. 2007. Protein stress and stress proteins: implications in aging and disease. J. Biosci. 32(3): 511–515.

Stephensen, E., J. Svavarsson, J. Sturve, G. Ericson, M. Adolfsson-Erici and L. Forlin. 2000. Biochemical indicators of pollution in shorthorn sculpin (*Myoxocephalus scorpius*) caught in four harbors on the southwest coast of Iceland. Aquat. Toxicol. 48. 431–442.

Stegeman, J.J. and P.J. Kloepper-Sams. 1987. Cytochrome P450 isozymes and monooxygenase activity in aquatic animals. Environmental Health Perspectives. 71: 87–95.

Stegeman, J.J. and J.J. Lech. 1991. Cytochrome P450 monooxygenase systems in aquatic species: carcinogen metabolism and biomarkers for carcinogen and pollutant exposure. Environmental Health Perspectives. 90: 101–109.

Stegeman, J.J. and M.E. Hahn. 1994. Biochemistry and molecular biology of monooxygenases: current perspectives on forms, functions and regulation of cytochrome P450 in aquatic species. In: Aquatic Toxicology: Molecular, Biochemical and Cellular Perspectives. Ed. D.C.Mallins, G.K. Ostrander, CRC Press, Boca Raton, FL. 87–206.

Stroka, Z. and J. Drastichova. 2004. Biochemical markers of aquatic environment contamination —cytochrome P450 in fish. A Review. Acta Vet. Brno. 73: 121–132.

Sun, Y., H. Yu, J. Zhang, Y. Yin, H. Shen, H. Liu and X. Wang. 2006. Bioaccumulation and antioxidant responses in goldfish *Carassius auratus* under HC Orange No 1 exposure. Ecotoxicological and Environmental Safety. 63: 430–437.

Svoboda, M., J. Kouril, P. Hamaskova, P. Kalab, I. Savina, Z. Svobodova and B. Vykusova. 2001. Biochemical profile of blood plasma of tech (*Tinca tinca* L.) during pre—and postspawning period. Acta Vet: Brno. 70: 259–268.

Tatara, C.P., M. Mulvey and M.C. Newman. 1999. Genetic and demographic responses of multiple generations of mosquitofish (*Gambusia holbrooki*) exposed to mercury. Environ. Toxicol Chem. 18: 840–2845.

Tierney, K., M. Casselman, S. Takeda, T. Farrell and C. Kennedy. 2007. The relationship between cholinesterase inhibition and two types of swimming performance in chlorpyrifos-exposed coho salmon (*Oncorhynchus kisutch*). Environ. Toxicol. Chem. 26(5): 998–1004.

Trant, J.M., S. Gavasso, J. Ackers, B.-Ch. Chung and A.R. Place. 2001. Development expression of cytochrome P450 aromatase genes (CYP19a and CYP19b) in zebrafish fry (*Danio rerio*). J. Exp. Zool. 290: 475–483.

Ugwemorubong, G.U., O.F. Gamaliel and O.D. Oveh. 2009. Enzymes in selected tissues of catfish hybrid exposed to aqueous extracts from *Lepidagathis alopecuroides* leaves. Electronic J. Environ. Agrical. Food Sci. 8(9): 856–864.

Vedel, G.R. and M.H. Depledge. 1995. Stress-70 levels in the gills of *Carcinus maenas* exposed to cooper. Marine Pollution Bull. 31(1-3): 84–86.

Vifia, J., F. Pallardo and C. Borras. 2003. Mitochondrial Theory of aging: importance to explain why females live longer than males. Antioxidants & Redox Signaling. 5: 549–556.

Vladimirov, Ju. A. and E.V. Proskurina. 2007. Lectures of Medical Biophysics. Moscow University Publ. «Academkniga». 432 pp (*in Russian*).

Vutukuru, S.S., N.A. Prabhath, M. Raghavender and A. Yerramilli. 2007. Effect of arsenic and chromium on the serum amino-transferaese activity in Indian Major Carp, *Labeo rohita* Int.J. Environ. Res. Public Health. 4(3): 224–227.

Wegwu, M.O. and S.I. Omeodu. 2010. Evaluation of selected biochemical indices in *Clarias gariepinus* exposed in aqueous extract of Nigerian crude oil (bonny light). J. Appl. Sci. Environ. Manage. 14(10): 77–81.

Weigand, C., S. Pflugmacher, A. Oberemm and Ch. Steiberg. 2000. Activity development of selected detoxification enzymes during the ontogenesis of the Zebrafish (*Danio rerio*). Int. Review of Hydrobiology. 85: 413–422.

WHO International Programme on Chemical Safety (IPCS). 1993. Biomarkers and risk assessment: concepts and principles. Environmental Health Criteria 155, World Health Organization, Geneva.

Widdows, J. and P. Donkin. 1991. Role of physiological energetics in ecotoxicology. Comp. Biochem Physiol. 100C: 69–75.

Winston, G.W. 1990. Physicochemical basis for free radical formation in cells: production and defences. In: Stress responses in plants: adaptation and acclimation mechanisms. R.C. Alscher and J.R. Cumming (eds.). Wiley, New York. 57–86.

Winston, G.W. 1991. Oxidants and antioxidants in aquatic organisms. Comp. Biochem. Physiol. 100C(1-2): 173–176.

Winston, G.W. and R.T. Di Giulio. 1991. Prooxidant and antioxidant mechanisms in aquatic organisms. Aquatic Toxicol. 19: 137–161.

Zelinski, S. and H.-O. Portner. 2000. Oxidative stress and antioxidative defense in cephalopods: a function of metabolic rate or age? Comp. Biochem. Physiol. 125B: 147–160.

Zowail, M.E.M., S.I. El-Deeb, S.S. El-Serafy, E.H. Rizkalla and H. El-Saied. 1994. Biochemical genetic studies of serum properties of family Mugillidae in two different habitats of Egyptian water. Bull. Natl. Inst. Oceanogr. Fish. Egypt. 20(1): 175–190.

CHAPTER 3

Development of Stress Biomarkers in Fish Embryos and Larvae

Fish display a wide range of developmental ontogenies. These distinctions have taxonomic, evolutionary, and ecological causes. The concept of saltatory ontogeny suggests that development is not gradual but proceeds in leaps separated by a series of stable developmental stages. In this context, endogenous/exogenous feeding also distinguishes the developmental phases of embryo (egg), eleutheroembryo (feeding off the yolk sac) and larvae (exogenous feeding) in fish (Belanger et al., 2010). Two fundamentally different strategies characterize the early development of marine teleosts (Pfeiler, 1986). The first and the most common strategy (type 1) consists of a post-hatch period in which the yolk-sac is resorbed; exogenous feeding begins immediately thereafter and it continues throughout the larval period as the larvae grow into juvenile fish. In the second strategy (type 2), after a similar post-hatch period during which the yolk-sac is resorbed, the larval fish shows a dramatic increase in size. Two potential sources of nutrition have been proposed for type 2 larvae: dissolved organic carbon and particulate organic carbon in the form of zooplankton fecal pellets and larvacean houses. Type 2 larvae may remain in the plankton for several months and are typical of five orders of bony fishes: the albuliformes (the bonefish), the anguilliformes (the eels), the elopiformes (the tarpon and ladyfish), the notacanthiformes (the spiny eels) and the saccopharyngiformes (the gulper eels). The type 2 larva has an unusual morphology. It is decidedly laterally compressed, almost leaf-like in appearance, with a perfectly clear body and a slender head that gives it its name: the leptocephalus. Development proceeds in two main phases. In phase I, the larvae grow in size until they reach a maximum that is typical of the species. During phase I, unlike most (type 1) fish larvae, energy reserves are accumulated within the leptocephalus as lipid and an acellular mass within a mucinous pouch. The mucinous pouch consists of proteoglycans, compounds made up of a conjugated peptide and glycosaminoglycan carbohydrates, most familiar as mucus and cartilage. Phase II of leptocephalus development consists of a size shrinkage and a profound change in shape to the juvenile morph, fueled by combustion of most of the accumulated energy reserves in the form of glycosaminoglycans and lipids (Pfeiler, 1996). The two elements in the energy budget equation that influence the amount of energy available for growth in larval fish are metabolic rate and excretion. Metabolic rate receives the greatest allocation of energy: 80–85% of the total ingested energy. Excretion commands a much lower percentage of ingested energy, with values ranging between 4–40% (Bishop and Torres, 1999). In early fish life the changes of metabolic rate, oxygen consumption and energetic metabolites lead to alterations in biomarker activities and concentrations.

Early fish developmental stages represent an attractive model for environmental risk assessment and the evaluation of pollution effects. Among them zebrafish is one of the excellent models for determination of fish development and the formation of biomarker mechanism during early life (Scholz et al., 2008). In diverse animal species maternally inherited mRNA and proteins are used to program the earliest stages of development but are degraded by the midblastula transition, allowing genetic control of development to pass to zygotically synthesis transcripts (Kishida and Callard, 2001). Biomarker development in early life of fish is very important for understanding their key role in the fish defense system and adaption to unfavorable environment.

3.1 Biotransformation System

3.1.1 Phase I

The cytochrome P450 monooxygenase system metabolizes a large number of xenobiotic compounds including many environmental pollutants. This metabolism can lead to detoxification, or in some cases, activation to reactive intermediates with toxic and carcinogenic effects. The members of this family are identified in early life stages of fish and other animals (Andersson and Förlin, 1992).

Cytochrome P4501A (CYP1A) activity is detectable at the stage of liver formation in embryos of killifish (*Fundulus heteroclitus*) and medaka (*Oryzias latipes*). The additional proteins are needed for CYP1A expression, are detectable in killifish and in zebrafish embryos at the onset of blood circulation (Powell et al., 2000; Andreasen et al., 2002). Thus, the cellular components of xenobiotic metabolism in fish are presented in early development (Incardona et al., 2005; Hornung et al., 2007).

Female zebrafish deposit mRNA into their oocytes (Hsu et al., 2002). Using *in situ* hybridization, they detect mRNA for the cholesterol side-chain cleavage enzyme, CYP11A1, in the embryos during epiboly and later in the yolk syncytial layer. CYP1A protein has also been detected in the trophoblastic syncytium in post-hatch larval sea bream *Sparus auratus* (Ortiz-Delgado and Sarasquete, 2004). It is possible that maternally derived P450 mRNA, such as the one encoded for CYP1A, is deposited within the medaka oocyte and then translated in the yolk or the yolk syncytial layer. Alternatively, CYP proteins may be deposited within the yolk during oogenesis and become localized or active in the yolk syncytial layer. An interesting possibility suggested by both alternatives is that induction of metabolic activity in female fish due to prior contaminant exposure may lead to higher levels of metabolism in embryos and larvae (Hornung et al., 2007).

CYP1A mRNA expressions of the three early embryonic stages of zebrafish were detected and kept at lower level throughout the whole three stages (Wu et al., 2008a). Expression for genes CYP1A, CYP 1B1, CYP1C1 and CYP 1C2 was studied during early embryogenesis of zebrafish (Jonsson et al., 2007). Each of the four CYP1 genes showed a characteristic time course of basal expression during zebrafish development. The first gene whose expression peaked was CYP1B1, which was maximally expressed within the first 2 days. The early expression in the developing embryo could be related to formation of the eye, as retina starts to develop at about 30 hpf (hours post fertilization) and vision begins around hatch in zebrafish. It could also be related to development of brain, heart, and other structures. CYP1B1 may contribute to retinoic acid synthesis during embryonic pattering, and hence may play a role in regulatory function during embryogenesis. The time for basal expression of the CYP1Cs differed from that of CYP1B1 in developing zebrafish. CYP1C1 and CYP1C2 transcript levels increased until around hatch (3–4 dpf) and then remained at this level for several weeks. The basal expression of CYP1C1 in killifish (*Fundulus heteroclitus*) embryos increased rapidly before hatch and then the increase slowed down. These findings suggest that CYP1C1 and 1C2 may play some physiological roles in newly hatched fish, perhaps related to transition from embryonic to free-swimming life. Then the basal expression of CYP1A increased several fold at hatching and continued to grow during the first week of development, peaking around day 21 at about 30-fold the 8hpf level. Hypoxia inhibited CYP1A activity and decreased its basal level measured by EROD activity, to 52% of normoxic control values in zebrafish larvae (Fleming and Di Giulio, 2011).

Cytochrome P450 aromatase (CYP19) gene expression regulates the ratio of androgens to estrogens and the expression of this enzyme is critical for reproduction and sex differentiation for most vertebrates including fish. Most vertebrates have a single CYP19 gene while zebrafish has two genes CYP19a and CYP19b. Both genes are expressed in gonads and in estrogenic extra-gonadal tissues (such as brain and pituitary). Both CYP19 genes are expressed in unfertilized eggs and whole embryo at the beginning of embryogenesis (1.5 hpf, 16 cells age) and decline progressively at 6 and 12 hpf. This pattern is consistent with transfer and degradation of maternally transcribed mRNA. An alternative possibility is that mRNA transcribed in granulose cells are transferred to oocytes by way of gap junction-like complexes that are present before ovulation or after the stage of follicle maturation in several fish species (Kishida and Callard, 2001).

A secondary rise of both isoforms was seen between 12 and 24 hpf, indicative of the onset of embryonic transcription; thereafter, stage-related patterns differed. An abrupt and dramatic, approximately 11-fold increase in relative signal intensity of CYP19b mRNA occurred between 24 and 48 hpf, and by 120 hpf (early larval period) the signal was about 25-fold higher than at 24 hpf (Kishida and Callard, 2001).

CYP19 transcripts were detected as early as 3 or 4 days post-fertilization (pdf) in developing zebrafish fry/juvenile at the period of ovarian (21 dpf) or testicular (25–28 dpf) differentiation, and peak abundance was detected on day five (Trant et al., 2001).

In addition, semi-quantitative analysis of CYP19 mRNA in gonads of Japanese flounder (*Paralichthys olivaceus*) during sex differentiation showed that there was no difference in the levels of CYP19 mRNA between the female and male groups when the gonads were sexually indifferent (50 dpf). However, after the initiation of sex differentiation (60 dpf) the mRNA levels increased rapidly in the female group, whereas they decreased slightly in the male group. The induction of sex reversal of genetically female larvae to phenotypic males by rearing them at high water temperature caused a suppression of CYP19 gene expression (Kitano et al., 1999).

CYP1A and their substrate EROD are the most studied enzymes in salmonids eggs and larvae. The CYP1A genes have been cloned from *Salvelinus*, *Salmo*, and *Oncoryhynchus* genera and indicate that at least two isoforms exist (Finn, 2007). Other genes expressed in response to xenobiotic exposure are the CYP3A isozymes, two of which have been cloned from rainbow trout. Embryonic and larval expression is restricted to the liver and foregut. However, the medaka studies demonstrated an ontogenic shift in the expression pattern of the two CYP3A isozymes. One isozyme (CYP3A40) was expressed early in development and continued throughout adult stages, while the second isozyme (CYP3A38) was expressed only after hatch. A similar finding has been reported for zebrafish CYP3A65 gene, which was initially expressed in the liver, but subsequently in the foregut following hatch (Finn, 2007).

At 1-week post-hatch stage medaka constant high EROD activity was observed in the liver and gills, while 4-week post-hatch see-through medaka exhibited less EROD activity than 2-week post-hatch see-through medaka.

1-day and 1-week post-hatch see-through medaka exhibited high intrinsic EROD activity in the liver, gills, and other organs. This intrinsic activity declined with growth and explained the high constant EROD activity at 1-week post-hatch stage (Kashiwada et al., 2007).

CYP1A1 was identified in cod eggs, larvae and juveniles and its level depends on developmental stage (Goksøyr et al., 1991). Biotransformation enzymes were studied *in vitro* in subcellular fractions from early life stages of Arctic charr (*Salvelinus alpinus*), whitefish (*Coregonus lavaretus*) and grayling (*Thymallus thymallus*). Each species showed an increase in 7-ethoxyresorufin O-deethylase (EROD) and NADPH-cytochrome reductase throughout embryolarval development. Significant increase in enzymatic activities seemed to occur around hatching period in the three species, especially for EROD activity. Interspecies comparison led to different results depending upon how activities were expressed (specific, per individual, or normalized with respect to the size of individual) (Monod et al., 1996). EROD activity was very low in sardine larvae caught in Bilbao estuary from the North Coast of Spain (Peters et al., 1994). Low level of CYT P450 activity was detected in the eggs of freshwater fish *Misgurnus fossilis* L. and the fluctuations of the enzyme level depended upon fish developmental stage (Isuev et al., 1991). Microsomal EROD activities in eyed stage rainbow trout embryos increased approximately 3 times from 3 to 7 days (from 0.25 ± 0.1 to 0.70 ± 0.1 pmol product/min/mg prot) (Koponen et al., 2000).

Therefore, the expression of CYT P450 genes was observed in early developmental stages of many fish species from various geographical locations and in laboratory conditions.

3.1.2 Phase II

As we have described previously (see Chapter 2.1.2) GSTs are the phase II enzymes which are involved in the biotransformation of various organic pollutants and endogenous substances. Activity of GST was identified in zebrafish embryos (Cazenave et al., 2006) and in larvae which was less than in adults tissue (Oliveira et al., 2009). At the same time in early development of the organisms, GST activity varied much more than in adults.

GST was identified in the larvae of zebrafish (Pauka et al., 2011). GSTs were studied *in vitro* in subcellular fractions from early life stages of Arctic charr (*Salvelinus alpinus*), whitefish (*Coregonus lavaretus*) and grayling (*Thymallus thymallus*). Each species showed an increase in glutathione S-transferase throughout embryolarval development. Significant increase in enzymatic activities seemed to occur around hatching period in the three species, especially for EROD activity. Interspecies comparison led to different results depending upon how activities were expressed (specific, per individual, or normalized with respect to the size of individual) (Monod et al., 1996).

The presence of mRNAs for metallothionein and glutathione S-transferase in eggs indicate that the potential for oxidative defense may be present in fertilised eggs. Genes for the enzymes non-specific carboxylesterase and phenol UDP-glucuronosyl transferase, which have roles in the metabolism of xendobiotic substances as well as pollutants, showed a peak in expression at hatch. CYP1A was expressed at hatch and the mRNA levels increased by about four-fold in 20 days post-hatch indicating the potential for metabolic activation of toxic polyaromatic compounds at the earliest free-living stages (Hodgson and George, 1998).

The activity of the detoxication enzyme GST differed during ontogenesis of zebrafish. The enzyme in all embryonic stages is very sensitive to toxic impact (Wiegand et al., 1999, 2000). Low GST activity ranged from 1.05 to 1.3 (nmol GST/min mg protein) was indicated in within 8 h of fertilization carp embryos (Palikova et al., 2007a,b). At the same time GST activity in zebrafish larvae was comparable with those in liver and head in adults, but it was greater than in gills and muscle (Coelho et al., 2011). Reduced glutathione (GSH) to 25 days post-hatch (dph) was indicated in the larvae of Asian Seabass, *Lates calcarifer.* GSH levels were low at gastrulation, indicating increased metabolic rate and formation of lipid radicals during this period, corresponding to the decrease in the level of ascorbic acid, which is consumed for regeneration of GSH (Kalaimani et al., 2008).

Glutathione-S-transferase (GST) was higher in eggs than in larvae of *Dentex dentex* (Mourente et al., 1999a). Overall, the activities of GST decreased in the early stages of development during a period where there was no exogenous dietary input into the larvae. Cytosolic GST activities in eyed stage rainbow trout embryos increased insignificantly from 3 to 7 days (from 159 ± 0.3 to 178 ± 1.5 nmol/min/mg prot.) (Koponen et al., 2000). During development of the larvae Japanese flounder the GSH concentration increased significantly in juveniles as compared to both larvae stages. GST activity was higher in settling larvae than in metamorphosing ones, but in juveniles tn was lower by more than 2-fold (Cao et al., 2010) (Table 3.1). Other researchers have observed the comparable values of GSH content and GST activity in Japanese flounder larvae are: 154.5 µg g^{-1} protein and 31.3 U mg^{-1} protein respectively (Huang et al., 2010).

GST activity increased progressively during early development of *Salmo iridaeus* and at the end of the embryogenesis it was 11-fold higher than those measured in the early stages. The GST fraction purified from fish embryos and from the adult tissues was identical (Aceto et al., 1994). On the other hand as for other aquatic animals the embryonic type of GST isoenzyme in the embryos of *Bufo bufo* falls at very low level in the adult liver (Falone

Table 3.1. Alterations of GSH level and GST activity in larvae and juveniles of Japanese flounder (adopted from Cao et al., 2010).

Developmental stages	GSH level, µg g^{-1} protein	GST activity, U mg^{-1} protein
Metamorphosing larvae (18 dph)	59.36	9.92
Settling larvae (33 dph)	58.71	37.15
Juvenile stages (78 dph)	98.09	15.48

et al. (2004). The authors proposed that during transition to adult life the animals leave the aquatic environment to live in the terrestrial locations, characterized by high oxygen concentration.

The results of several researchers demonstrated that the cytochrome GSH level and GST activity are present in early developmental stages of fish, but their fluctuations during early life are not identical and the data obtained is contradictory, depending upon fish species, developmental stages and experimental duration.

3.2 Lipid Peroxidation and Antioxidant System

Antioxidant enzymes play a key role in inactivation of reactive oxygen species (ROS) and thereby control oxidative stress as well as redox signaling. Both processes change across the life span of the organism and thus modulate its sensitivity and resistance against free radical damage (Vifia et al., 2003; Wegwood et al., 2011; Hu et al., 2007).

During development, fish embryos and larvae are sensitive to the environmental factors-both endogenous and exogenous. Oxygen availability is vital for embryo development where cell division and *de novo* tissue formation require increased metabolic rates and stimulation of oxygen consumption in early developmental stages. Increased uptake of exogenous oxygen may have the potential to affect pro-oxidant processes in fish early life. During embryogenesis, the metabolic energy substances change between glucose, free amino acids and lipids depending upon species and developmental stage and these metabolic fuels may provide endogenous sources of free radical production and the alteration of antioxidant status of the organism (Peters and Livingstone, 1996).

3.2.1 Lipid peroxidation

Generation of reactive oxygen species (ROS) was detected even in the unfertilized egg of devil stinger by chemiluminescence analysis. ROS was continuously detected during the development from fertilized egg to larvae and tended to increase gradually. ROS was produced on the surface of embryo and the head region of larvae, especially peripheries of eyes. It suggested the presence of NADPH oxidase-like ROS generating system in the embryo of devil stinger which is already activated at fairly early stage of development before the maturation of usual immune system (Kadomura et al., 2006).

Lipid peroxidation level measured as lipid peroxide and peroxidizable lipids in embryos and 3-day larvae turbot *Scophthalmus maximus* demonstrated a 13-fold increase in lipid peroxide after hatching and was mirrored by an indicated 3-8-fold decrease in peroxidizable lipid (Peters and Livingstone, 1996). LPO level in metamorphosing larvae of Japanese flounder was determined as 2.67 nmol MDA mg^{-1} protein, in settling larvae it was greater (5.41 nmol MDA mg^{-1} protein) and in juveniles it dropped again to 3.1 nmol MDA mg^{-1} protein (Cao et al., 2010). MDA content in the Japanese flounder larvae was detected as 2.7 nmol MDA mg^{-1} protein (Huang et al., 2010).

The level of lipid peroxidation (LPO) measured by malonic dialdehyde (MDA) was very high in fertilized eggs of *Dentex dentex* prior to hatching and it was significantly higher than in larvae which was decreased during development (Mourente et al., 1999a). The ratio of MDA to polyunsaturated fatty acids PUFA in the eggs was greater than 5, whereas the ratio in larvae was much lower and generally increased from approximately 1.6 in newly-hatched larvae to 2.4 in day 9 larvae. Thus, the eggs contained a relatively high level of MDA, but upon hatching the larval level of MDA was only about one-third of the present in the unhatched egg. The authors suggested that two-thirds of the MDA was not associated with the developing larvae body but rather may have been associated with the chorion and/or perivitelline fluid. It may reflect the accumulation of MDA during embryonic development with excretion from the developing embryo into the perivitelline fluid which was lost upon hatching. Subsequently, MDA levels decreased over the whole time-course of larval development reflecting decreased accumulation through enhanced antioxidant enzyme activities and/or increased excretion rate (Mourente et al., 1999b). Increase of lipid peroxidation from 10 dph to 25 dph during larval development of Asian seabass (*Lates calcarifer*) was also documented (Kalaimani et al., 2008). Enhanced LPO only occurred at the metamorphosis stage (days 19 to 28) of *Solea senegalis*. This could be due to auto-oxidation of the PUFA as a consequence of increased oxygen presence due to higher metabolism during this stage (Sole et al., 2004).

Our studies have shown the fluctuations of lipid peroxidation compounds in Black Sea fish embryos and larvae. The dynamics of LPO through fish embryogenesis was not uniform. We studied the trends of prooxidant-antioxidant processes during the early development of Black Sea round goby (Fig. 3.1) which is highly distributed in fish species and in which early life is well known (Fig. 3.2).

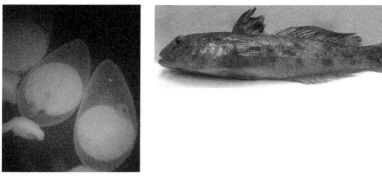

Fig. 3.1. Developing embryos and adult fish of round goby *N. melanostomus.*
Color image of this figure appears in the color plate section at the end of the book.

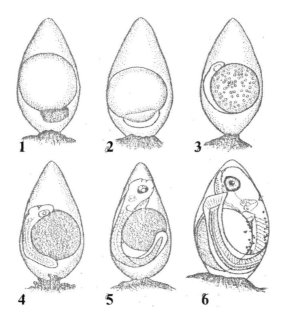

Fig. 3.2. Embryonic development of *N. melanostomus* 1-morula; 2-gastrulation; 3-formation of embryonic fascia; 4-segmentation of tail; 5-eyes pigmentation; 6-embryo before hatching (Kalinina, 1976).

Lipid oxidation factor in *N. melanostomus* was the highest in stage III and then decreased during embryonic development and in larvae. In *P. marmoratus* and *P. maxima* this factor varied less and tended to decrease to the end of fish embryogenesis (Fig. 3.3). Similar trends were observed for ketodiens and dien conjugats levels. The highest concentration of TBA-reactive products was indicated in stage V for all examined fish species

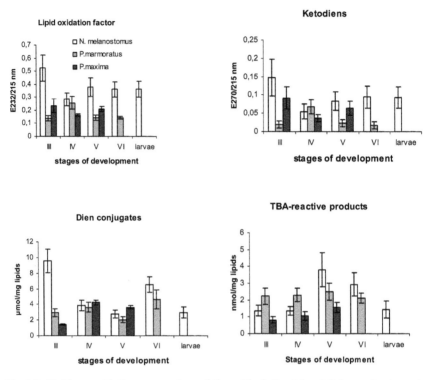

Fig 3.3. Lipid peroxidation parameters of three Black Sea teleost fish species in early developmental stages (mean ± SEM). III-VI–stages of embryogenesis.

and then it decreased to the end of embryogenesis and in larvae of *N. melanostomus*. High level of LPO was detected at the periods of intensive growth and organogenesis of the embryos required high consumption of oxygen and energy and used the different substrates for organ formation.

Fluorescent spectra of lipids characterize the level of the end products of LPO or the level of fluorescent age pigments (FAP). Our measurments showed the highest value of lipids fluorescence at the wave of 356 nm. At stage VI of Gobiidae embryo the level of fluorescent compounds increased more than 2–3 fold and in larvae of *N. melanostomus* the maximum was detected at 356–370 nm and the value increased in 2-fold as compared to eggs at stage VI ($p < 0.01$) (Fig. 3.4). FAP level could reflect larval metabolic activity and its fluctuations depend on nutritional deprivation in larvae (Mourente et al., 1999b). Fish peroxidation status may modify by the diets: low in vitamin E showing significantly higher values of thiobarbituric acid reactive compounds (TBARC) (Mourente et al., 2000).

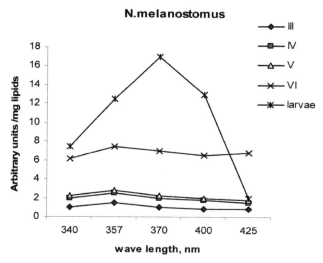

Fig. 3.4. Fluorescent spectra of lipids of fish eggs and larvae in Black Sea round goby *N. melanostomus.* (n = 3–5). I–VI-stages of embryonic development.

Therefore, lipid peroxidation processes vary during fish embryogenesis and hatching larvae and they depend on the growth, organogenesis, stage of development, oxygen consumption and the type of eggs (benthic or pelagic).

3.2.2 Antioxidant enzymes

Antioxidant enzyme activities are indicated in developing eggs of marine animals. Ovoperoxidase was detected in the eggs and it uses hydrogen peroxide as substrate which is generated in NADPH-oxidase system on the surface of egg shell. Superoxide anion and other substances containing tyrosine cation are formed during the process of egg development, then they degrade by ovoperoxidase which protects the eggs against ROS damage (Takahashi et al., 1989; Winston, 1991).

Antioxidant enzyme CAT and SOD activities were similar in larvae of *Sardina pilchardus* and were estimated as 4 µmol min^{-1} mg protein and 7 SOD units min^{-1} mg protein respectively. The activities of both antioxidant enzymes were similar to those observed for 8–12 mm larvae of sprat *Sprattus sprattus* and 3 day old turbot *Scophthalmus maximus* (Peters et al., 1994, 2001).

The trends of antioxidant enzyme activities in three early developmental stages of *S. maximus* were not uniform. SOD activity was determined to be higher in embryos than larval stages, while CAT and GR activities increased during the development from embryo to the free swimming larvae (11-days post-hatch). It may be indicative of a decreasing need to detoxify O_2^- during development from embryo to 11-day-old larvae. Following fertilization, the oxygen uptake by fish embryos increases and peaks after the embryo hatches. During *S. maxima* embryo development, rates of oxygen consumption increased 3-fold from fertilization to hatching and then it reached the maximum between 3–4 days post hatch. The increasing rate of oxygen uptake may increase pro-oxidant processes such as O_2^- production, which in *S. maximus* counteracted by SOD activity present in the embryos and larvae.

Oxygen availability is one of the factors effecting hatching. Hypoxia or agents that cause hypoxia (low sea water dissolved concentration of cyanide or high concentrations of dissolved hydrogen), stimulate hatching whereas hyperoxia may suppress hatching. If this phenomenon is physiologically relevant to the hatching of *S. maximus* the embryonic SOD may act to reduce elevated tissue concentration of O_2^- which may occur during respiratory chain and/or trans-membrane electron transfer during hatching. A simultaneous increase in CAT, GR and SeGPX activities with development of embryos to larvae would indicate a progressive need to remove H_2O_2 and lipid peroxides from tissues. Increasing of GR and SeGPX activities with respect to larval developmental stages may indicate an increased requirement for glutathione both for SeGPX activity and as free radical scavenger for radical derived from lipid peroxidation and oxidation of other substances. SeGPX is not the only source of larval GPX activity which has been associated with isoforms of the GST. Regulation of GST expression in fish may occur via factors other than ROS production. NADPH-dependent DT-diaphorase activities in larvae were lower than in embryos and reflected a decrease potential for ROS generation via redox cycling endogenous or dietary compounds (quinones). DT-diaphorase is an important enzyme in vitamin K synthesis and it is possible that during *S. maxima* larval development, vitamin K-dependent processes, i.e., synthesis and utilization may affect larval DT-diaphorase activity (Peters and Livingstone, 1996).

The variations of antioxidant enzymes level were documented in early developmental stages of marine fish *Dentex dentex* (Mourente et al., 1999a,b). After hatching, activities of CAT and SOD varied differentially with CAT generally increasing and SOD generally decreasing during early larval development. The reason of this is perhaps not immediately apparent as SOD, by converting superoxide to hydrogen peroxide, directly provides substrate for CAT. The activity of GPx was initially low in unhatched eggs, but like CAT, it increased during larval development and the largest value occurred between days 3 and 4 post-hatch, just after the activity of CAT also showed its greatest increase between days 2 and 3 post-hatch. Therefore, this growth of enzymes activity occurred before yolk sac absorption (day 5 post-hatch) and disappearance of the oil globule (day 6) and, in addition, there was no correlation between these increases in enzyme activities and fatty acid composition, which declined steadily. The increase levels of CAT and GPx may be in response to other metabolic changes occurring at this time possibly leading to change in sources of oxidative stress. The activity of GR was the lowest activity of all the enzymes. The imbalance in GPx and GR activities could potentially lead to GSSG accumulation and a lower GSH/GSSG ratio (Mourente et al., 1999b).

Changes in SOD and CAT activities were shown in early development of Japanese flounder (Table 3.2). SOD activity decreased more than 2-fold in metamorphosing and settling larvae as compared to juveniles; CAT activity varied less (Cao et al., 2010). In an other study of larvae of Japanese flounder development the similar values of the enzyme activities were observed: 66.9 U mg g^{-1} protein for SOD, 21.9 U mg g^{-1} protein for CAT (Huang et al., 2010).

We observed that during embryogenesis of several Black Sea species most of the enzymes tended to increase in eggs and especially in hatching larvae while the levels of low molecular scavengers decreased (Rudneva, 1999) (Fig. 3.5).

Our findings noted the interspecies differences between three benthic species. The enzyme activities tended to increase during embryogenesis and especially in larvae in the case of SOD and CAT in all examined species (Rudneva, 1994, 1995, 1997, 1999). Our results agree with the data of other investigators who reported the similar trends of antioxidant enzyme levels in fish early development. For instance, the catalytic activity tended to increase during *Solea senegalensis* larvae development and significant changes were seen in most enzymes. The main changes in enzymatic activities (CAT, t-GPx, GST, DT-D), occurs at 3 dph, when sole larvae starve, as reserves are finished and from this point they initiate feeding with rotifers. Important enzymatic changes have also been reported in other fish species when the metabolic energy source in larvae changes from endogenous lipids to exogenous food. At this stage the general metabolic rate could be enhanced, and consequently, more damage to proteins could occur (Sole et al., 2004).

Table 3.2. Alterations of SOD and CAT activity in larvae and juveniles of Japanese flounder (adopted from Cao and et al., 2010).

Developmental stages	SOD activity, U mg g^{-1} protein	CAT activity, U mg^{-1} protein
Metamorphosing larvae (18 dph)	50.20	24.76
Settling larvae (33 dph)	21.71	15.92
Juvenile stages (78 dph)	24.76	23.00

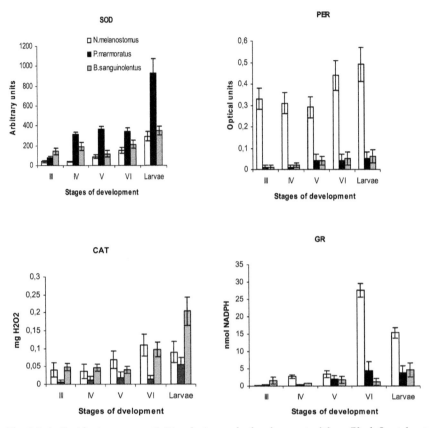

Fig. 3.5. Antioxidant enzyme activities during early development of three Black Sea teleost fish species: gobies *Neogobius melanostomus, Proterorhinus marmoratus* and blenny *Blennius sanguinolentus* (mg⁻¹ protein min⁻¹, mean ± SEM).

Among the H_2O_2-removing systems in *Salmo iridaeus* embryo development, CAT seems to be the only enzyme present at significant level and in this case CAT plays more important role in preventing H_2O_2 damage in fish embryo as compared to GPX. During further development activity of antioxidant enzymes SOD, CAT, GR, GLX increased maximum at stage 33 when the formation of the swim bladder as well as pigmentation had occurred. At the end of development the activity of glutathione-containing enzymes increased progressively and values were found to be about 85–312 fold higher than in early stages, while the variations of CAT and SOD activities were lower (Aceto et al., 1994).

Three antioxidant enzymes, SOD, CAT and GPX, showed high activities during gastrulation, suggesting an increased metabolic rate during the period of embryonic development of *L. calcarifer*. Though the SOD activity apparently decreased progressively during 3–20 dph of larval development, the difference was not significant. CAT showed high activity during gastrulation and remained constant up to 3 dph, suggesting an increased need to metabolise hydrogen peroxide (H_2O_2) and organic peroxides. In contrast, SeGPX activity increased progressively from 5 dph to 25 dph during larval development, indicating an increased need to detoxify lipid peroxides. This is evident from the observation of increased lipid peroxidation from 10 dph to 25 dph during larval development (Kalaimani et al., 2008). The altertions of low molecular weight antioxidants during fish early life are presented below.

3.2.3 Low molecular weight scavengers

We detected low molecular weight antioxidants in early developmental stages of Black Sea teleosts and elasmobranchs. Eggs of Black Sea elasmobranch *Squalus acanthias* contain high level of antioxidants: 9.80 ± 3.30 µg% of glutathione, 6.71 ± 1.22 mg% of vitamin K, 0.45 ± 0.02 mg g^{-1} lipids of vitamin A, 1.4 ± 0.12 µg g^{-1} lipids of carotenoids, and 0.70 ± 0.13 mg g^{-1} wet weight of vitamin E. Antioxidant levels in Black Sea teleost *Gobiidae* embryos and larvae are presented in Table 3.3.

Vitamin A concentration decreased in hatching larvae as compared to developing embryos. Fluctuations of vitamin A content in embryos were not significant in *N. melanostomus*. In *P. marmoratus* the levels of vitamin A and carotenoids decreased progressively at the end of the embryogenesis.

Ascorbic acid (AA) plays an important role in fish gametogenesis because gonads development stimulated gonadotropin associated with the direct interaction between katecholamines and steroid hormones and their receptors regulate the mechanism of ascorbate absorption, its transport and metabolism (degradation and synthesis) in reproductive system. Ascorbic acid is one of the main components of fish diet which plays an important

Table 3.3. Low molecular weight scavengers in early developmental stages of Black Sea *Gobiidae* (n = 3, mean ± SEM) (Rudneva, 1999).

Antioxidants	Developmental stages				
	III	IV	V	VI	Hatching larvae
Neogobius melanostomus (n = 3)					
Vitamin A,mg/g lipids	2.19 ± 0.36	1.77 ± 0.38	2.75 ± 0.81	2.91 ± 0.67	0.98 ± 0.05
Proterorhinus marmoratus (n = 3)					
Carotenoids, µg/g lipids	38.84 ± 8.16	26.76 ± 5.32	10.52 ± 2.35	14.13 ± 3.67	
Vitamin A,mg/g lipids	1.96 ± 0.38	0.61 ± 0.15	0.20 ± 0.12	0.38 ± 0.12	

role in the success of reproduction because of: 1. its antioxidant function in gametes and protection DNA in them against endogenous metabolites and exogenous substances after hatching (exogenous stress), 2. its consumption from food for the improvement of the spawning processes, 3. participation of AA in the endocrine regulation of neurohormonal—hormonal pathway in fish reproduction and 4. participation of AA in regulation of maturation and/or sterility at the high doses of UV-radiation or in the intensive aquaculture conditions (hypoxia, hyperoxia, food pro- and antioxidants) (Dabrowski and Ciereszko, 2001).

At the period of early development of Atlantic halibut level AA varied in eggs and larvae and depended on its content in the food. The levels of AA and vitamin E were constant during the development of yolk sac and was observed as 170 and 131 ng per larvae correspondingly (Ronnestad et al., 2005). In new hatching larvae the values decreased to 80 and 97%. At the period of larvae development the content of AA and tocopherols differed from each other. At the beginning of feeding (approximately 200 days after hatching) more than 95% of AA and lower than 30% of vitamin E of the yolk was transferred to larvae body. The consumption of vitamin E in larvae body came entirely from yolk absorption. In eggs of rainbow trout (AA) level decreased to 50% at day 3 after fertilization and in yolk sac the level was estimated as 1/3 from the value of those in non-fertilized egg (Otto and Moon, 1996).

Ascorbate dynamics study in embryos and larvae of Sea bass (*D. labrax*) and Sea bream (*Sparus aurata*) demonstrated the similar trends (Terova et al., 1998). In Sea bass newly fertilized eggs (stage I) obtained from the broodstock receiving vitamin C reach diet, the mean concentration of ascorbic acid (AA) was $218.5 \pm 17.7 \ \mu g \ g^{-1}$ wet weight and then decreased significantly from stages I–V. The AA concentration at stage VI (144 h after spawning and just before feeding) was $173.2 \pm 19.2 \ \mu g \ g^{-1}$ wet weight. Sea bass not receiving the addition of vitamin C diet AA content in newly fertilized eggs (stage I) was significantly lower compared to supplemented group ($155.9 \pm 6.9 \ \mu g \ g^{-1}$ wet weight) and again decreased through stages II and III, then increased during hatching stage (IV).

In Sea bream AA concentration of the newly fertilized eggs (stage I) was $122.4 \pm 5.1 \ \mu g \ g^{-1}$ wet weight in the supplemented group and it remained almost constant through stage III, increasing significantly at stage IV. The following significant increase in larval stages led to $173.8 \pm 13.7 \ \mu g \ g^{-1}$ wet weight at stage VI, corresponding to 96 h after hatching. In the group fed, the diet without supplementation AA concentration was $103.9 \pm 3.5 \ \mu g \ g^{-1}$ wet weight, just after fertilization. It maintained almost constant concentration through stage III and increased significantly at stage IV, following thus the same dynamics as the supplemented group (Terova et al., 1998). Ascorbic acid was evaluated during the period from gastrulation (GS) to 25 days post-hatch (dph) in the larvae of Asian Sea bass, *Lates calcarifer* (Kalaimani et al., 2008).

Hence, Sea bass broodstocks fed with a diet supplemented by AA, produced eggs with a concentration of the total ascorbic acid in the newly fertilized egg stage significantly higher compared to the non-supplemented AA broodstock. The concentration of AA increased between the third and the fourth stage in correspondence with an increase in the dry matter. The loss of chorion and perivitelline fluid during the hatching stage, which are considered as parts of the egg not containing vitamin C assumed to be located in the embryo and yolk, may explain the apparent differences of the total ascorbate concentration in the hatched larvae of both species. The depletion of yolk-sac reserves of vitamin C to meet the nutritional demands of the larvae before first feeding may be a reasonable explanation of it. The increase of AA observed after the third stage in all groups may suggest the existence of a reducing environment at the hatching stage, transmitting dehydroascorbic acid (DAA) in AA, a more biologically active form, in order to meet the demands of the newly hatched larvae (Terova et al., 1998). The collapse of vitamin E level in the developing larvae of *Dentex dentex* between 7 and 9 post-hatch indicated that tocopherol was being consumed rapidly at this stage in the process of quenching free radicals and chain-breaking. It only occurred after yolk-sac resorption and disappearance of the oil globule. These processes may be related as there may be transient increase in PUFA into the metabolically active pool at this time due to resorption of yolk PUFA and that cannot be incorporated into membranes in starving animals and therefore must be oxidized (Mourente et al., 1999a).

High level of vitamin A was detected in the eggs of Sea bass *Dicentrarchus labrax* before and after fertilization, in developing embryos and hatching larvae, then it significantly decreased. During larvae development the level of vitamin E decreased slowly at the period of 4 days and then progressively increased from 9 to 40 days. In abnormal larvae vitamin E content was significantly higher than in normal ones (Guerriero et al., 2004).

Development of fish larvae *Salmo salar* depended upon vitamin A content in eggs. The eggs were grouped in to two classes: high level of vitamin A (3.3 ± 0.1 µg retinol per g dry weight) and low level of vitamin A (2.2 ± 0.3 µg retinol per g dry weight). Before fertilization the eggs were incubated at temperature 14°C and 8°C during the period of 133 days. At high temperature the level of vitamin A increased progressively in embryos and larvae after hatching as compared to the case of low temperature. On the other hand vitamin A content in larvae tissues and temperature did not influence on fish fry (Ornsrud et al., 2004).

SH-containing substances such as glutathione and amino acids play an important role in eggs fertilization and the further development. The direct correlation between the success of eggs fertilization and embryos survival and the level of SH-groups in protein and non-protein substances was documented because free thiol groups on the egg surface react with growth factors (Konovalov, 1984). GSH participates in various critical cellular processes including detoxification and the regulation of cellular proliferation and development. Also, GSH has been reported as a co-factor in thiol-disulfide exchange reactions in sea urchin eggs and in the protection of protein-SH groups. These groups are involved in cell division, and their oxidation results in damage to this important function. In fact, the progress of embryonic development is delayed by oxidative stress and protein-thiol group oxidation (Ahmad, 2005).

Therefore, early developmental stages of fish contain enzymes which metabolize superoxide anion (SOD), H_2O_2 (CAT and SeGPX), detoxify organic peroxides (SeGPX and total PER) and reduce the potential of redox cycling by endogenous and foreign compounds (DT-diaphorase). The enzyme activity trends are differed during embryogenesis and in larval stages in different fish species and reflect the peculiarities of fish early life and the sources of oxidative stress. Our results suggested that in early stages of embryonic development of marine fish the low molecular weight antioxidants play an important protective role against oxygen damage. During embryogenesis the concentration of low molecular weight scavengers decreased, while an enhancement of antioxidant enzyme activities was detected, possibly as a result of gene expression. Probably, in hatching larvae, the antioxidant enzymes are the main oxygen protective mechanisms. During the next period of life cycle larvae feed extensively on microalgae rich in carotenoids, vitamins and other antioxidants. Like other marine animals, examined fish species mobilize carotenoids and tocopherols from the microalgae and accumulate them in the body, especially in eggs and gonads to protect them against oxidative stress.

3.3 Stress Proteins

Stress proteins such as Hsp70 and chaperonins (Hsp 60) families and metallothioneins play a role in protein homeostasis, protecting the cell from proteotoxicity and transporting damaged proteins to the lysosomes for degradation. They are expressed from the early larval stages of development in fish although their profile and/or induction potential may vary according to the species, developmental stage, protein forms studied and culture. Their interest in stress monitoring also lies in the proven link between Hsp response and disease in fish (Sole et al., 2004).

During vitellogenesis, many exogeneous and endogenous materials are transferred from the mother fish to the oocytes and subsequently to eggs and yolk sac larvae. MT mRNAs, and MTs are transported to oocytes of various organisms (Mommsen and Walsh, 1988). In fact, maternal MT mRNAs are found in the oocytes and unfertilized eggs of two species of sea urchins, rainbow trout (Olsson et al., 1990) and other animals. MT may be transferred to offspring in Mozambique tilapia (Lin et al., 2000) and zebrafish (*Danio rerio*) (Riggio et al., 2003) through a maternal effect. The role of MT during vitellogenesis is to induce vitellogenin and to regulate essential metals (like Zn and Cu) in lake trout (*Salvelinus namaycush*) eggs (Werner et al., 2003).

Low levels of MT mRNA were found in newly fertilized tilapia eggs. The fertilized eggs of rainbow trout had the lowest MT mRNA levels among the developmental stages examined in this species (Olsson et al., 1990). It is unlikely that the maternal MT mRNA observed in developing oocytes had been degraded. Alternatively, the tertiary structure of maternal MT mRNA may have changed and did not hybridize with cDNA probe. In order to extend their half-lives, many maternal mRNAs go through structural changes and remain dormant during embryo development. This state is also viewed as translational regulation of maternal mRNA. By binding with ribonucleoproteins (RNPs), these non-translated mRNAs are masked from the general translational apparatus. No attempt was made to remove these RNPs. The low-molecular-weight MT mRNA was possibly extrinsically taken up along with other maternal materials into the oocytes. If this is the case, offspring with a higher MT content would be from parental females containing more MT (Lin et al., 2000).

The amount of MT mRNA was very low in all early stages in tilapia *Oreochromis mossambicus* oocytes, fertilized eggs and newly hatched larvae and their alterations through the early development were insignificant. The level in oocytes within ovaries → newly fertilized eggs → newly hatched larvae were the following: 3.93 ± 1.14, 5.25 ± 2.50 and 3.20 ± 2.11 100 MT µg per total loading protein µg correspondingly (Lin et al., 2000). The MT level tended to increase in larvae development of tilapia from pre- hatched larvae HO-stage (438.1 ± 31.61–465.4 ± 31.78 ng/mg protein) to newly hatched larvae (261.5 ± 57.59–383.0 ± 49.19 ng/mg protein) and 1 day post hatch larvae (539.3 ± 128.69–605.7 ± 62.09 ng/mg protein) (Wu et al., 2008).

Zebrafish (*Danio rerio*, Hamilton 1822, Pisces, Cypriniformes) are popular test organisms in developmental genetics as well as in ecotoxicology. Stress proteins Hsp70 were found in eggs and larvae of the fish (Scheil et al., 2009).

The presence of mRNAs for metallothionein in plaice eggs indicates that the potential for oxidative defense may be present in fertilised eggs (Hodgson and George, 1998). MT contents were identified in 3-day old larvae of tilapia (*Oreochromis mossambicus*) at the level of 150–200 ng mg^{-1} protein (Wu et al., 2007).

Stress proteins were present in all stages of benthic fish, *Solea senegalensis* at the period of 28 days from hatching, showing distinctiveness in the molecular weight of Hsp60 at hatching day at 1 dph. Quantitatively, both forms were significantly elevated at 3 dph, with their ratio (Hsp70/Hsp60) decreasing over time (Sole et al., 2004). The highest values were found in the endogenous phase (0–2 dph) and the lowest value at 3 dph, before the external food was initiated. A steady-state situation was displayed from 6 to 28 dph and it was related to a high demand for metals in developmental stages. This behavior could be similar in the initial stages of larvae, where the external input to support energy requirements is limited. The steady-state of MT levels reached after the initiation of external feeding could also have an antioxidant role associated with the oxyradical scavenging capacity of MT.

During stressful conditions, Hsps are enhanced to repair damaged proteins but they are also able to confer adaptation to further environmental insults. As far as stress proteins are concerned, two bands in developing larvae of *S. senegalensis* were detected by immunoblotting, corresponding to the Hsp70 and Hsp60 forms with a molecular weight of approximately 78 and 66 kDa, respectively. The molecular weight of the band 66 kDa showed an increase from hatching day to 2 dph in all blots when it remained stable. No changes in this protein's weight or conformation have previously been reported at early stages of development. A feasible hypothesis could be that the chaperoning functions of this protein within the cell are different at particular larval stages, when they bind to different polypeptide chains. Only the end of the endogenous phase seemed to be a sufficiently stressful period to register in the stress proteins profile (Sole et al., 2004).

The variations of stress proteins during larval development has not been reported in many fish species. The study on their trends (Hsp70 and 90) was followed in silver sea bream (*Sparus sarba*) from 1 to 46 dph, showing insignificant fluctuations in the stress protein content during the first 14 dph and significant increase thereafter (Deane and Woo, 2003). Hsp70 protein burden was not significantly changed in the early life stages of *Salmo trutta f. fario*, under controlled conditions (Luckenbach et al. 2003).

No major differences were seen in the protein content of Hsp70 and Hsp60 at the early larval stages of *S. senegalensis*. However, the Hsp70/Hsp60 ratio decreased with development, suggesting a major role of the Hsp60 form as the animal grows. This is supported by the fact that Hsp60 is better represented in adult fish. Nevertheless, Hsps play a key role during development, such as myogenesis, in addition to their traditional "housekeeping" task. However, in the case of Hsp60 at hatching day and 1 dph, they were slightly lower, 62 and 64 kDa, respectively. From a quantitative perspective the highest Hsp70 and Hsp60 contents were displayed at 3 dph, whereas the lowest corresponded to 28 dph for Hsp70 and hatching day for Hsp60. It is worth noting that the Hsp70/Hsp60 ratio decreased with age throughout the 28-day study period, with a parallel evolution during development. As for the total stress protein mean content (Hsp 60 + 70), it was similar at all stages, averaging between 9.6 and 9.8 a.u/mg prot but was enhanced at 3 dph, at which it was 16.0 a.u/mg prot (Sole et al., 2004).

Hence, stress proteins and metallothioneins play a role in protein homeostasis, protecting the cell from proteotoxicity and transporting damaged proteins to the lysosomes for degradation. They are expressed from the early stages of fish development and their level varies during early life depending on maternal status, fish species specificity and environmental conditions.

3.4 Biochemical Parameters

Growth performance is linked to digestive or energetic capacities in the early life stages of a salmonid species. The comparative study of two strains (Fraser and Yukon gold) of Arctic charr are known to have different growth potentials during their early development. Trypsin, lipase, and amylase activities of whole alevins were measured at regular intervals from hatching through 65 days of development. To assess catabolic ability, the measurements of five enzymes representing the following metabolic pathways: amino acid oxidation (amino aspartate transferase), fatty acid oxidation (β-hydroxy acyl CoA-dehydrogenase), tricarboxylic acid cycle (citrate synthase), glycolysis (pyruvate kinase), and anaerobic glycolysis (lactate dehydrogenase) are used. The measurement of these enzyme activities in individual fish allowed a clear evaluation of digestive capacity in relation to energetic demand. Higher growth performance appears to be linked to lower metabolic capacity (Lemieux el al., 2003). The fluctuations of biochemical composition and enzyme activities were observed during the early development of Pacific halibut, *Hippoglossus stenolepis* (Whyte et al., 1993).

Changes in subcellular localization of alanine and aspartate aminotransferases and their activities were found at different pH optima in each subcellular fraction of the rainbow trout *Parasalmo mykiss* L. and chum salmon *Oncorhynchus keta* during ontogenesis (Samsonova et al., 2005). Activity of LDH in zebrafish larvae was less in 2-fold than in adult tissues (Oliveira et al., 2009).

Lactate dehydrogenase (LDH) is the terminal enzyme in anaerobic glycolysis in vertebrates and it is an indicator of anaerobic potential. LDH activities in red drum *Sciaenops ocellatus*, and lane snapper, *Lutjanus synagris* were approximately 10-fold lower than values typical for adult fish; LDH and citrate synthase (CS) activities increased during early developmental stages, but nutritional effects were apparent. Clear differences (up to 4-fold) between well-fed and starving fish were evident in both LDH and CS activity in red drum. Differences between well-fed and poorly fed larvae were evident until 9 d after hatching. Lane snapper larvae reared at 25°C had significantly lower LDH activities than larvae reared at 28°C (Clarke et al. 1992).

The isozymic forms of four dehydrogenases, separated by disc electrophoresis in acrylamide gels, have been studied during development and in adult organs in the cyprinodontiform fish, *Oryzias latipes*, the medaka. Xanthine dehydrogenase appears as a single molecular form at hatching, and in the adult fish it is found only in the liver and gut. Glucose-6-phosphate dehydrogenase is a single enzyme in early development, just before hatching two new forms appear, and after hatching all three forms decrease in whole larval extracts. In adults, the enzymes were found only in brain, liver, and ovary. Malate dehydrogenase increases from a single molecular form in early stages to five or six after hatching. Each adult tissue has its characteristic assemblage of isozymes. The only lactate dehydrogenase detectable before hatching is isozyme 5, and four others appear abruptly at hatching. All adult organs have one to three LDH isozymes, but isozymes 1 and 2 were found only in the retina of the eye. The abrupt change from the single isozyme 5 to the full complement also occurs in eyes of the larvae taken pre- and post-hatching. Homogenates of pre-hatching embryos do not inhibit the electrophoretic separation or reactivity of isozymes 1, 2, 3, and 4 of post-hatching larvae. Presumably hatching triggers the appearance of the more rapidly migrating isozymes, especially strongly in the retina. On the basis of the structure of LDH isozymes in other organisms, it can be concluded that polypeptide A is present before hatching, but some restraint on the synthesis or activation of polypeptide B seems to be removed by hatching. The importance of hatching at the time of qualitative change in these four enzymes is correlated with increase in other metabolic activities at this time (Nakano and Whiteley, 1965).

Using electrophoresis, the patterns of LDH and G-6-PDH isoenzymes were identified in the early embryonic and larval stages of *Ctenopharyngodon idellus* until the 14th day post-hatching. LDH isoenzymes (A4, A3B, A2B2, AB3 and B4) detected in such stages were found to be controlled by two loci, LDH-A and LDH-B. Such isoenzymes in control individuals exhibited variable activities during development. From 8-hour-post-fertilization (8h-stage) to 24h-stages, only the maternal B4-isoenzyme was detected. Such enzyme in addition to other maternal metabolic ones is necessary to initiate the embryonic development. The amount of maternal B4-isoenzyme fluctuates until the 24h-stage, in which it decreases to 11%. LDH-B locus and other enzymes involved in glycolysis and in carbohydrate metabolism are important as "housekeepers" in developing embryos and almost remain constant or decrease slowly. Generally, such situations refer to the utilization of maternal enzymes stores and their subsequent degradation before zygotic translation of mRNA. 16h-stage up to 36h-stage is the time of gastrulation and epiboly during which the body axis formation and organogensis are initiated. This developmental period corresponds to the initial differentiation of tissues of liver, kidney, gonads, nervous system, etc. The zygotic genes of *Ctenopharyngodon idellus* embryo appear to be inactive up to the process of organogensis. Only, in 36h-stage (the hatching stage), a major switching process from maternally determined LDH translation to embryonic (zygotic) LDH translation occurred. The zygotic LDH-isoenzymes varied in their individual activities and their number from stage to stage (Mekkawy and Lashein, 2003).

From the 8 h- to 16 h-stages, only the maternal G-6-PDH-A isoenzymes were detected. The amount of G-6-PDH in these stages fluctuated and was influenced by allele A2 or A3. The contribution of different alleles in the subsequent stages varies. However, that of A2 was higher in the majority of larval stages (Mekkawy and Lashein, 2003).

Most of the egg carbohydrates are associated with the egg membrane and therefore are probably unavailable for use by the developing embryo at least until hatching; the time of their release from that membrane. However, intensive catabolism of carbohydrates commences at fertilization indicating that carbohydrates play an important nutritive role during initial cleavage. Such very early metabolism of carbohydrates is associated with LDH and G-6-PDH isoenzymes. (Mekkawy and Lashein, 2003).

Citrate synthase, located in the mitochondria and positioned at the beginning of the Krebs citric acid cycle, catalyzes the formation of citrate from acetyl-CoA and oxaloacetate. Citrate synthase activity correlates directly with oxygen consumption rate in fish and is a good index of maximum aerobic potential. Herring *Clupea harengus* and plaice *Pleuronectes platessa* were reared at 8 and 12°C from the fertilized egg to a larval age of up to 600 degree-days. Soluble protein as well as the activities of both citrate synthetase (CS) and lactate dehydrogenase (LDH) were measured in homogenate supernatants of individual larvae at 10°C. All scaling factors showed significant differences between the species. Within species, the scaling factors for CS activity were either small or not significantly different between the two rearing temperatures, but the scaling factors for the LDH activities were significantly different at the two temperatures for both species. Herring larvae, which had higher LDH activities when newly hatched, showed smaller scaling factors for LDH ($b = 1.42$ at 8°C and $b = 1.07$ at 12°C) than plaice ($b = 2.11$ at 8°C and $b = 1.45$ at 12°C). Activities converged as the larvae grew. The results of the current study together with reanalysis of data from the literature indicate an increasing aerobic and anaerobic capacity during the larval stage of fish (positive allometry) (Overnell and Batty, 2000).

The comparative analysis of LDH activity in Atlantic menhaden *Brevoortia tyrannus* in laboratory-reared and wild populations was provided. Larvae of whose first feeding was delayed had depressions in protein-specific LDH activities, and LDH activity appeared to scale with length. Wild larvae had LDH activities within the range of activities found in laboratory-reared larvae and were classified into 3 nutritional categories based on length and LDH activities. Fewer than 30% of larvae collected from the majority of stations were classified in superior condition. No differences were detected between larvae collected during the day and those collected at night (Fiedler et al., 1998).

The study of ontogeny of the digestive functions in the fertilized eggs and yolk sac stages of the *Sparus aurata* larvae showed that the amylase and tyripsin activity were detected in fertilized eggs. Leucine alanine (LEU-ALA) peptidase activity was higher than other enzymatic activities throughout 96 h after hatching. Aminopeptidase N(LAP) activity decreased from fertilized egg to hatching. After hatching, LAP activity increased until 24th hour and then decreased up to the beginning of exogenous feeding (P < 0.05). The lowest alkaline phosphatase activity (AP) was taken from fertilized eggs. Then, AP activity tended to increase until the end of experimental period (P < 0.05). The findings demonstrated the nutritional requirements of seabream larvae at the start of exogenous feeding (Naz, 2009).

RNA/DNA ratio in larval fish is an indicator of both condition and growth rate. Since larvae grows exponentially, condition and growth rate are closely associated. The researchers have shown that the RNA/DNA ratio is very sensitive to changes in feeding levels in larval winter flounder *Pseudopleuronectes americanus* and larval cod *Gadus morhua*. A positive linear relation between RNA/DNA ratio and growth rate was observed at 5, 7, and 10°C. (Buckley, 1982).

The data suggested that the majority of the main enzymes and metabolites which involve in biosynthetic and energetic processes through the early development are present in oocytes, fertilized eggs and developing embryos. Some of them are house keeperes and the others are transferred from maternal organism to the embryo via blood stream or by pinocytosis. During early development of embryos and larvae the ratio of different components and enzyme activities is changed which depends on gene expression and metabolic rate in embryos and unfed larvae. In feeding larvae these parameters fluctuate depending upon food composition and its availability.

3.5 Metabolic Rate

The rate of physiological processes in developing embryo is reflected in the metabolic rate and it depends upon the individual characteristics such as size, developmental phase, nutritional status and activity (Pakkasmaa et al., 2006). The energy lost in excretion can be divided into two subcomponents: feces, the portion of ingested energy that is indigestible, and nonfecal nitrogen, which in fish is usually in the form of ammonia or urea. Teleosts are ammoniotelic, excreting the majority of their nitrogen in the form of ammonia (Bishop and Torres, 1999).

At the end of the embryogenesis the moving activity, respiration rate and heart systole are increased and that also requires high energy costs (Ozernuk, 1985). The metabolic rate in fish embryos also depends upon the energetic substrates (carbohydrates, lipids) and their changes in different stages of the early development which is associated with yolk sac consumption, specificity of morphogenesis and growth (Zhu et al., 1997; Desvilettes et al., 1997). Lipids and fatty acids (FAs) play an important role in successful early development of marine fish both as cataboic substrate and as structural components of cellular membranes. In addition, investigations during the last decade have shown that free amino acids (FAA) abound in marine pelagic fish eggs. The concentration in the egg (about 150 mM) is well above what is typically found in adult teleostean tissues. FAA have been implicated as a fuel in the energy metabolism of developing marine fish eggs and larvae. Pelagic eggs of marine teleosts generally have a high content of FAA. The large pool of FAA is almost exclusively contained within the yolk sac compartment. The FAA pool is depleted during development and reaches low levels at first feeding. FAA seem to be utilized to varying extents for body protein synthesis, but are more often used as substrates in energy metabolism (Rønnestad and Fyhn, 1993).

Pacific cod embryos are capable of synthesizing lipids and FAs from other energetic reserves during development. This increase of lipid occurs late in the egg stage and is accompanied by reduction in egg density shortly before hatch, suggesting that lipogenesis may be important for provisioning buoyant energetic reserves to larvae as they transition from the demersal to pelagic region. Demersal eggs have been considered to have very little energetic reserve comprised FAA. For example, the FAA component represents 20–50% of the total amino acids (TAA) in pelagic eggs, whereas it represents only 4% in demersal eggs. There is a large component of non-lipid organic content in the egg, which is likely FAA and available both as substrate for basal metabolic demands and as the energetic source for lipogenesis prior to hatch. The cost of converting these additional energetic reserves to lipid is unknown, but it is undoubtedly less efficient than catabolizing lipid directly from maternal contributions. In Atlantic halibut the calories associated with the 28% lipid increase accounted for the majority (73%) of the total calories associated with the decrease of FAA from fertilization and hatching (Laurel et al., 2010).

Lipids is a major energy source during the early development period in *Dentex dentex* (Mourente and Vazquez, 1996). Lipids content in fish eggs steadily decrease with development until hatch, after which there is often a more precipitous rate of lipid loss in free-swimming embryos. The eggs were grouped by two classes based on lipid composition. The first class "low lipid eggs" contain < 5% wet weight of total lipids and high concentration of polar lipids (> 60%, mainly phospholipids such as phosphatidylcholine). The second class "high lipid eggs" contain > 5% wet weight dominated by neutral lipids such as triacylglicerids (TAGs) (Tocher, 2003). In anchoveta eggs FAA and protein together accounted for 48% of the suitable substrates for embryo development in batches II and IV. In fish eggs without oil drops (such as anchoveta eggs) FAA and proteins meet at least half of the energy demands in egg development. During development of egg from stage I–III, 65–69% consumption of yolk (batches I and II) and rapid consumption of FAA were observed as well as losses of 50% of the lipid and 24% of TAG content were observed (Krautz et al., 2010).

Our findings demonstrated that at the end of embryogenesis and in hatching larvae the level of triacylglycerides declined and the level of free fatty acids grew in *Gobiidae* species. Triacylglycerides are the main energetic source in egg yolk and in some cases the success of larvae hatching and their further survival depend on these components concentration (Mourente et al., 1999a,b). During the early life the total lipid content decreased in the eggs of aquatic organisms, which directly correlated with the decline of triacylglycerides concentration. Lipid changes showed a permanent use of the endogenous reserve and no lipid synthesis was detected in any developmental stage (Roche-Meyzaud et al., 1998). With this connection it was suggested that triacylglycerides may involve not only in the energy generation, but they are also used for phospholipids production, the main structural components of cell membranes in developing embryos. At the same time the changes in sterols/phospholipids ratio demonstrated the modifications in cell membrane fluidity in larvae as compared to the embryos which was documented by the other investigators also (Evans et al., 1997). The interspecies peculiarities in lipid composition of fish were also found and they depend on yolk chemical composition, female feeding and food consumption (Moskalkova, 1984, 1985).

Both in teleosts and elasmobranchs lipids are the main components of the egg. Mature pre-ovulated ovarian follicles of 10 australian chondrichthyans were high in lipid content, indicative of a large energetic expenditure and high maternal investment. Larger lipid reserves were found in viviparous dogshark (28–36% wet weight, ww) compared to oviparous chimaeras (19–34% ww) and catshark (18% ww). Neutral lipids and monounsaturated fatty acids were the main source of lipidic energy during vitellogenesis and gestation. For most species, there was a peak in total lipid content, levels of storage lipids and essential fatty acids at time of ovulation (Pethybridge et al., 2011). In our studies we observed that lipids were the main components of the eggs of Black Sea shark *Squalus acanthias* and the eggs of benthic fish species.

Among teleost fish species there is unusual correlation between growth and energy consumption. Study on the metabolic processes of leptocephalus larvae demonstrated the coefficients relating mass-specific metabolic processes to mass through direct determination of metabolic rate, excretion rate and intermediary metabolic enzyme activities. The mechanism responsible for the unusual relationship between mass and metabolic rate in these larvae is the formation of an energy depot in the form of glycosaminoglycans. The larvae increase rapidly in mass but accumulate little metabolizing tissue, thereby maintaining low overall metabolic costs in very large larvae (Bishop and Torres, 1999).

Throughout embryogenesis the eggs use different substrates for oxidation and energy generation. For instance, *Scophthalmus maximus* embryos demonstrated glycogen dependence following the first 18–19 h, free amino acids (84%) together with a small amount of phosphatidyl choline (9%) and later wax esters (5%) comprised the metabolic fuels of embryonic development. Following hatch (day 4.4 post fertilization) wax esters (33%) and triacylglycerols (25%) were initially catabolised with the remaining free amino acids (10%) (Finn et al., 1996). We have recorded the fluctuations of lipid components level in developing embryos of *Psetta maxima maeotica*. The decrease of diglycerides and fatty acids level during embryo development from stage II to stage IV was observed while the concentration of triglycerides varied insignificantly. Wax level also dropped from the stage II to stage IV. We could conclude that neutral lipids and wax catabolised in the early phases of embryogenesis for energy generation and for cell division (Finn et al., 1996). Thus the decrease of lipid content especially neutral origin (triacylglycerides) during early development of aquatic organisms is the general trend and our data have suggested this opinion. However, lipid composition and their ratio showed interspecies differences which depend of biological and ecological peculiarities on the examined species.

As we described previously (see Chapter 2.5) heat dissipation measured by microcalorimetry is the preferable method of the metabolic rate trends analysis during fish early life. Our findings observed the growth of heat production to the end of the embryogenesis of fish and especially in hatching larvae (Fig. 3.6). Hatching time was similar in both fish species. The metabolic rate increased constantly throughout the early development of both species and especially in hatching larvae. The reason of this fact lies in the active movement of the embryo for going out from the egg's shell (Rudneva and Shaida, 2000, 2006, 2012).

The similar trends were noted in developing pelagic eggs and larvae of Black Sea turbot *P. maxima maeotica* (Fig. 3.7) (Rudneva and Shaida, 2000; Rudneva et al., 2001, 2004). However, in pelagic turbot eggs which characterized short embryogenesis (4–5 days) the highest level of heat

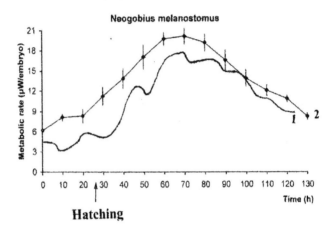

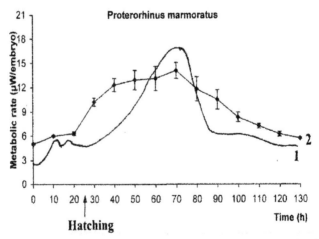

Fig. 3.6. Calorimetric tracings (1) and mean rates of heat dissipation (2) of developing eggs and larvae of two Black Sea *Gobiidae*. Vertical bars indicate ± SD, n = 4. The integral output signal was recorded once per 10 minutes, temperature of the incubation is ± 20°C, time of the monitoring is 130 h (Rudneva and Shaida, 2012).

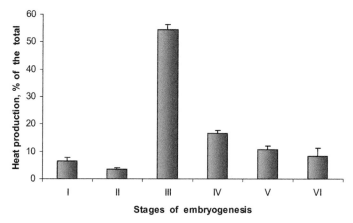

Fig. 3.7. Heat production of developing eggs of *P. maxima maeotica* in different periods of embryogenesis. Temperature of the incubation is +16,6°C, time of the monitoring is 80 h.

dissipation was detected in the stage III embryo (54.5% of the total heat production), at the period of intensive growth and morphogenesis. At the end of the embryogenesis metabolic rate in this species remained stable before hatching. Similar changes in heat production throughout early life of aquatic organisms were demonstrated by other investigators (Baker and Mann, 1991; Canepa et al., 1997; Geubtner, 1998; Schmolz et al., 1999; McCollum et al., 2005; Pakkasmaa et al., 2006). Heat production fluctuations in early life of terrestrial invertebrates were also documented (Penttinen and Holopainen, 1995; Lamprecht, 1998; Russel et al., 2009).

The obtained results demonstrated close link between the stages of the embryogenesis and the values of metabolic rate of the developing embryos and hatching larvae. We could conclude that the stage III was the longest and the important period of the turbot embryogenesis which characterized cell divisions, formation of the main organs, associated with the high metabolic activity and increase of energy consumption, respiration and oxidation. During the other periods of the embryonic development (stage IV–V) the embryo grew progressively and the metabolic rate was relatively stable and then dropped at the end of the embryogenesis (stage VI).

Our study demonstrated that metabolic rate in *Gobiidae* benthic eggs and larvae was similar while the heat production of pelagic developing embryos of *P. maxima maeotica* differed significantly. Peculiarities of the fish eggs' size, ecological status and yolk sac composition closely associated with the period of embryonic development: in pelagic eggs it is much more shorter than in benthic eggs. Fish eggs contain different energetic resources which involve in the metabolism of the embryo in different phases of development and yolk dependence. Thus the changes of the main energetic substrates such as lipids level may reflect the quality and quantity of developing processes in yolk-dependent fish embryos belonging to different ecological groups (benthic and pelagic). Lipids are the main components of yolk and the important source of the energy for developing embryo and their trends throughout embryogenesis are correlated closely with the total metabolic rate of the organism.

Besides that the metabolic rate of the individuals demonstrated high variability and closely depends upon both the environmental factors and genetic status. For instance it was shown that the metabolic rates in Arctic charr *Salvelinus alpinus* eggs vary between families and eggs from different families can be both genetically and maternally mediated (Pakkasmaa et al., 2006). Other investigators documented increase of metabolic rate in feeding Atlantic cod *Gadus morhua* larvae as compared to unfed individuals (McCollum et al., 2006). Several data of metabolic rate during early development of the animals are presented in Table 3.4.

On the other hand, resource availability may influence the cost of high metabolic rates which was demonstrated in eggs of fish species belonging to different ecological groups—benthic and pelagic. In both cases the increase of metabolic rate of developing fish embryos through early life is a general

Table 3.4. Metabolic rate of some fish species in early development.

Fish species	Developmental stage	Metabolic rate	References
Salvelinus alpinus	Eyed-stage egg	2.3–7.9 µW ind^{-1}	Pakkasmaa et al., 2006
		0.06–0.22 µW mg^{-1}	
	Hatching larvae	16.7 µW ind^{-1}	
	Hatching larvae	0.67 µW mg^{-1}	McCollum et al., 2006
Gadus morhua	30dph larvae	2.14 µW ind^{-1}(unfed)	
		16.56 µW ind^{-1}(fed)	
		12.6 µW ind^{-1}(unfed)	Geubtner, 1998
		21.7 µW ind^{-1}(fed)	
Neogobius melanostomus	Egg, stage V–VI	3.70 µW ind^{-1}	Rudneva and Shaida, 2012
	Hatching larvae	14.30 µW ind^{-1}	
Proterorchinus marmoratus	Egg, stage V–VI	3.28 µW ind^{-1}	Rudneva and Shaida, 2012
	Hatching larvae	12.12 µW ind^{-1}	
Psetta maxima maeotica	Egg, stage I–VI	1.2–5.3 µW ind^{-1}	Rudneva and Shaida, 2012
	Hatching larvae	4.8 µW ind^{-1}	
Atherina hepsetus	1-day larvae	5.2 µW ind^{-1}	Rudneva and Shaida, 2000
Atherina mochon pontica	Larvae	36.6 µW ind^{-1}	Rudneva et al., 2004; Rudneva and Shaida, 2006
Lepadogaster lepadogaster lepadogaster	Pre-larvae	1.63 µW ind^{-1}	Kuzminova, 2003

tendency in various fish species which is probably caused by increase of oxygen consumption and respiration rate, heart systole, moving activity and stress of hatching, when embryo goes out into environment.

Thus, trends of heat production together with the changes of energetic substrates composition could reflect the metabolic rate and correlation between energy generation and utilization in developing fish embryos and larvae. These results could be used for evaluation of critical periods of embryogenesis identification, hatching successes and larvae survival which is important for aquaculture and ecotoxicology purposes. However, several metabolic processes are affected by some chemical and physical factors which may change metabolic rate.

3.6 Cholinesterases

ChE inhibition has been widely used in aquatic ecosystems as a biomarker for organophosphorus pesticides (OP) exposure in aquatic organisms. However, the biological, environmental, and methodological factors affecting ChE activity have not been well documented and must be considered and understood before ChE activity can be used as an OP dependable indicator of exposure to aquatic organisms. Early life stage assays also contributed towards providing relevant information regarding anomalies in larvae development and behavior and the formation of response on unfavorable conditions and neurotoxic substances (inhibited cholinesterase). Thus, it is very important to study the development of ChE activity and its fluctuations in fish early life.

Several studies analyzing ChE activity in zebrafish embryos indicate that this enzyme is very important in neuronal and muscular development and its alteration may be associated to effects like embryo abnormal posture and spine malformations (Oliveira et al., 2009).

ChE activity was identified in larval stages of walleye (*Stizostedion vitreum*) and the activity of the enzyme depended upon developmental stage (Phillips et al., 2002a). Acetylcholinesterase activity has been detected in early medaka embryos including the yolk syncytial layer (Fluck, 1982). Activity of ChE in zebrafish larvae was more than 2-fold less than in adult tissues (Oliveira et al., 2009) and it was less sensitive to toxic impact in eggs than in adults and larvae (Domingues et al., 2010).

The study of the influence of water temperature, size of larval and juvenile walleye (*Stizostedion vitreum*), stress, long-term storage, postmortem changes, and methods of euthanasia on ChE activity demonstrated that all these factors modified enzyme activities in fish larvae. Water temperature (17.2, 20.9, and 24.6°C), stress, long-term storage (up to 180 days), postmortem changes, and method of euthanasia had no effect on ChE activity of walleye. There was a strong positive correlation

($r = 0.87$) between whole body ChE activity and total length (7.2–17.9 mm) for larval walleye, but a negative correlation between brain ChE activity and total length (59–164 mm) for juvenile walleye ($r = 0.75$). Size, age and development may affect ChE activity, fish of similar size should be used when evaluating the effects of ChE inhibitors. If fish of similar size are not available, it is recommended that relations between size, age, and development be understood so that estimates of variation in ChE activity can be made (Phillips et al., 2002a,b).

Cholinesterase and carboxylesterase activities have been measured in larvae of gilthead seabream, *Sparus aurata*, during the endogenous feeding stage, and ChE was characterized with the aid of diagnostic substrates and inhibitors. The results of the study showed that whole-body ChE of yolk-sac seabream larvae possesses typical properties of acetylcholinesterase (AChE) with an apparent affinity constant (K(m)) of 0.163+/−0.008 mM and a maximum velocity (V(max)) of 332.7+/−2.8 nmol/min/mg protein (Arufe et al., 2007).

During the period of development from stage 34 embryos to 24-h-old larvae of medaka *Oryzias latipes*, AChE activity rose rapidly, and this pattern appears similar to that reported for rainbow trout (Uesugi and Yamazoe, 1964). Histochemical staining of acetylcholinesterase activity appears in neural tissue along with heavy staining of skeletal musculature. It is possible that in early life stage of teleosts the rapid development of the cholinergic system fosters a dependence on this system, for example neurotransmission controlling gill movement during respiration, that results in higher sensitivity to the lethal effects of anticholinesterase pesticides (Hamm et al., 2001).

Therefore, ChE was detected in early developmental stages of fish in eggs and larvae. They were used as good diagnostic tools of fish development and growth. They are present in musculature and brain of the larvae and their activity is changed through stages of development.

3.7 Endocrinology of Fish Early Development

Development is epigenetic, which refers to changes in gene activity during development that are mediated by environmental (chemical) signals. Autocrine, paracrine (such as growth factors), and endocrine (such as steroid) signals coordinate the direction of differentiation of cells and tissues during critical periods of development. The formation of organs thus involves a complex cascade of signals whose action is dependent on being released at precise time and within a specific dose range. Coordination of these processes depends upon the translation of genes coding for these signaling molecules and their reception to appropriate times and appropriate rates. Organ systems responsive to the sex steroids include the male and female reproductive organs, the central nervous system, and the immune system, whereas thyroid hormone affects most of the times (Bigsby et al., 1999).

In most fish species the information about sexual development during early life is very limited. In some cyprinid fish the gonad may remain undifferentiated for a considerable time after hatching (e.g., 90 days in carp) and almost nothing is known about endocrinology of these early life stages. There is no information about measurable differences in the titres of sex hormones, or Vtg, between genetic males and females. Preliminary observations obtained in fathead minnow indicated that at 30 days posthatch (the time of sacrifice) gonads in fish were sexually differentiated if they were female (histologically, an ovary could be seen), but this was not necessarily the case for the putative males. Males could easily be distinguished from females by the difference in the concentration of Vtg in whole body homogenates. The presence of Vtg in females at a stage of sexual development well before the start of uptake of Vtg into the ovary is a phenomenon found in other fish species (Tyler et al., 1999). Low concentrations of Vtg were found in early development of anchoveta (Krautz et al., 2010) and juvenile chub (Zlabek et al., 2009). During vitellogenesis, many exogenous and endogenous materials are transferred from the mother fish to the oocytes and subsequently to the eggs and yolk sac of larvae (Mommsen and Walsh, 1988; Selman and Wallace, 1989; Scheil et al., 2009; Wiegand et al., 1999). The major vitellogenic steps included the exogenous step that is through the gonadotrophin to stimulate estrogen activity and to induce vitellogenin from hepatic to oocyte (Wu et al., 2008).

Fish reproduction depends on complex factors both exogenous and endogenous. In response to environmental signals such as temperature and photoperiod integrated through the brain, the hypothalamus secretes gonadotropin-releasing hormone (GnRH). GnRH stimulates the release of gonadotropins (GtH) from the pituitary, where GtHs and ovarian steroid hormones regulate oocyte growth and maturation. GtH secretion is regulated through a feedback mechanism by sex steroids. Two forms of GtH have been isolated from: GtHI and GtH II which are analogous to mammalian follicle stimulating hormone (FSH) and luteinizing hormone (LH), respectively. GtH I is involved in gametogenesis and steroidogenesis, whereas GtH II is involved in the final maturation stages of gametogenesis. Generally the gonadotropins are responsible for stimulating the synthesis of sex steroids (androgens and progestines) which in turn act on target tissues to regulate gametogenesis, reproduction, sexual phenotype and behavioral characteristics. The development of oogenesis is controlled by GtH in the majority of fish. Circulation of GtH I increase during early oocyte development and bind within receptors on the thecal and graulosa cell layer of the follicle. The thecal cells synthesize testosterone, and a low aromatization to result in the formation of estradiol in the granulose layer prior to plasma secretion. Subsequently, plasma estradiol will bind to estrogen receptor (ER) and trigger the series of steps resulting in the production of vitellogenin, and eggshell zona radiates proteins in the liver. Vtg and Zrp are released from the liver into the blood and are incorporated into the oocyte, through receptor-mediated endocytosis. As the oocyte continues to grow, levels of GtH I begin to decrease and are replaced by increasing levels of GtH II. The cascade reactions axis that leads to the production of maturation proteins is known as hypothalamus-pituitary-gonadal-liver-axis (HPGL-axis), which is ultimately controlled by feedback systems. Alterations in steroid hormone production could ultimately affect the feedback pathways, which could lead to impairment of reproductive process (Arukwe, 2001).

In male fish, GtH I is elevated throughout spermatogenesis and decreased at time of spawning whereas GtH II is typically low throughout the growth process and is elevated at spawning. The gonadotropins stimulate the proliferation of spermatogonia as well as synthesis of androgens required for gametogenesis and development of secondary sex characteristics. Androgen synthesis usually takes place in the Leydig cells. The type of androgen synthesized is dependent upon the species and developmental stage but may include testosterone, 11-ketotestosterone and/or androstenedion. A decline of androgen levels and a sharp increase in progestins during spawning are primarily attributed to elevated levels of GtH II (Arukwe, 2001).

17β-Estradiol (E_2) is the major estrogen in female teleosts, but large amounts of the androgen, testosterone, are also produced by the ovary. In the ovarian two-cell model, the theca cells synthesize testosterone, which is subsequently aromatized by cytochrome P-450 aromatase (CYP19) to E_2 by the granulose cells. E_2 stimulates the production of vitellogenin (Vtg) and eggshell Zr protein by the liver of female fish. The mechanism involves binding of the ligand (E_2) to estrogen receptors (ER) which are subsequently activated to form ER homodimers. These ligans-ER dimmer complexes initiate transcription of target genes by binding to estrogen response elements (EREs) in the upstream regulatory region of the genes. (Arukwe, 2001; Goksoyr, 2006).

During oogenesis of oviparous fishes, vitellogenin, a complex phospholipo-glycoprotein with molecular weights ranging from 300–600 kDa, is taken up into the oocyte by micropinocytosis (Mommsen and Walsh, 1988; Selman and Wallace, 1989) and lipids, carotenoids, glycoproteins, vitamin-binding proteins, hormones and RNAs are transferred to oocytes via the blood stream (Mommsen and Walsh, 1988). These factors are important not only in energy supply for metabolism and enzyme activity, but also in translational control of oocyte mRNA after fertilization.

Vtg and Zrp are proteins required for normal oocyte maturation in the developing female fish.Vtg is a bulk (molecular mass 250–600 kDa) and complex cadmium-binding phospholipoglycoprotein produced in the liver, secreted in the blood and transported to the oocyte where it is incorporated as constituents of the yolk. Zrp (zona pellucida protein, vitelline envelopeprotein, VEP) is the structure constituting the zona radiate (zona pellucida, vitelline envelope) of the eggshell. The Zrps are also produced in the liver as two or three monomers, and transported to the ovary for cross-linking into the eggshell. Vtg and Zrp are both regulated by estrogens in the maturing female fish. Vtg is an inherently unstable protein designed for degradation after being taken by the oocyte (Arukwe, 2001; Arukwe and Goksoyr, 2003).

Oocyte growth in fish is due to the uptake of systemic circulating Vtg, which is then modified by, and deposited as yolk in the oocyte. Vtg is selectively sequestered by growing ovarian follicles by receptor-mediated endocytosis before deposition in the oocyte. These specific oocyte Vtg receptors are clustered in clathrin-coated pits. Coated vesicles fuse with golgian lysosomes in the outer ovoplasm, of the oocytes and form multivesicular bodies. The golgian lysosomes contain cathepsin D, which processes Vtg into yolk proteins. Vtg is an important source of nutrients for egg and larvae, making the vitellogenesis an important developmental process. In addition, teleost eggs contain maternal sex steroids, cortisol, and other lipophilic hormones like thyroxin that may enter the egg through Vtg. These hormones may act as metabolites or as synergists with other substances during early life (Arukwe and Goksoyr, 2003). The yolk of an ovulated egg consists largely of material that have been deposited in the growing oocyte to be later utilized by the nascent embryo as a source of nutrition and energy to support development (Goksoyr, 2006).

All Vtg proteins have the following characteristics: they are female-specific serum or plasma proteins, precursors to yolk proteins, induced by estrogen, lipoglicophosphoproteins with molecular masses ranging from 300 kDa to 600 kDa and carrier proteins with both lipid and ionic component (e.g., calcium, zinc, cadmium, iron, etc.). There are three classes of Vtgs–VtgA, VtgB and VtgC which differ from each other by specific structure and corresponding genes.The Vtg A and VtgB are differently incorporated into oocytes at the ratio of 2 to 3 (Goksoyr, 2006).

The envelope surrounding is interface between the egg and sperm and between embryo and its environment. The egg envelope is a major structural determinant of the eggshell in fish, and is often referred to as zona radiata because of its striated appearance under the light of microscope. In fish the egg envelope is much thicker than in mammals, providing physical protection from the environment and playing a role in diffusive exchange of gases. The micropyle is closed within minutes after the eggs are activated by exposure to fresh water, which initiates a cortical reaction necessary for development of fertilized eggs. Ionic concentration of the medium lower than 0.1 M is needed for complete activation. After activation the zona radiate takes up water, gains resistance to breakage and can support up to 100 times more weight than oviductal eggs. The synthetic site of Zrp is the liver in most teleost species, but in some fish species Zrp synthesis appears in the ovary. Zonagenesis is the E_2-induced hepatic process of eggshell proteins, their secretion and transport in blood to the ovary and uptake into maturing oocytes (Arukwe and Goksoyr, 2003).

Therefore, fish reproduction is controlled by environmental signals which stimulate the endocrine system reactions and cascade of processes of hormone synthesis, oocyte maturation and spawning process. Oocyte growth in fish is due to the uptake of systemic circulating Vtg, which is then modified by, and deposited as yolk in the oocyte. Plasma estradiol binds to estrogen receptor (ER) and triggers the series of steps resulting in the production of vitellogenin and eggshell zona radiate proteins in the liver. Vtg and Zrp are released from the liver into the blood and are incorporated into the oocyte, through receptor-mediated endocytosis. Vtg and Zrp are proteins required for normal oocyte maturation in the developing female fish. The normal development of fish embryos depends upon the complex of all these factors and the destruction of only one of them may damage the norma; embryogenesis, hatching and further larvae survival and growth.

3.8 Development of the Immune System in Fish

Immune cells in fish development are present in Fig. 3.8. Although similarities are striking, there are also significant differences between fish and mammals with regard to the employment of different antibody classes. Homologous genes of many transcription factors involved in mammalian hematopoiesis have been used to trace corresponding development in fish by *in situ* hybridization. However, knowledge on the development of the immune system in fish is fragmented and limited compared to higher vertebrates. In some species, such as angelfish, initial hematopoiesis appears in yolk sac blood islands. In other species it appears in the intra-embryonic intermediate cell mass (ICM), as in zebrafish, or first in the yolk sac and later in the ICM, as in rainbow trout. Definitive hematopoiesis is established in the kidney, but an intermediate, larval site of hematopoiesis has been revealed in the tail of zebrafish (Hordvik, 2009).

Adaptive immunity in fish is based on molecular components with overall similarity to the mammalian counterparts, i.e., immunoglobulins, T cell receptors and MHC antigens, and a large number of accessory molecules (Sullivan and Kim, 2008). As in mammals, the T cells appear to be of either alpha/beta or gamma/delta type (Nam et al., 2003), and

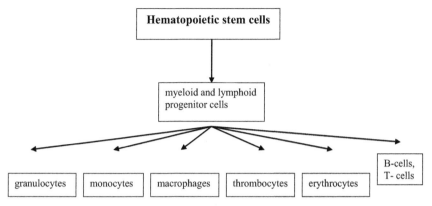

Fig. 3.8. Development of immune cells in fish (adopted from Hordvic, 2009).

the T cell receptor is in a complex with CD3 molecules (Yun et al., 2008). Leukocytes in fish comprise lymphocytes, thrombocytes, granulocytes, monocytes and macrophages. In contrast to mammals, thrombocytes and erythrocytes in teleost fish are nucleated (Hordvik, 2009). In fish, clusters of melanomacrophages are believed to form the teleost analogue to the germinal centres of mammalian lymph nodes, where specialized antigen-presenting dendritic cells interact with T cells, thereby initiating adaptive immune responses. It is assumed that melanogenesis is an important immune mechanism in fish (Haugarvoll et al., 2006). This feature seems to have been lost in mammals.

In recent years *Danio rerio* is used as a good organism for study of the immune system in early development. Definitive hematopoiesis is established in the kidney, but an intermediate, larval site of hematopoiesis has been revealed in the tail of zebrafish. Early differentiation of zebrafish macrophages has been shown before the onset of blood circulation. These primitive macrophages can phagocytose and kill bacteria injected in the zebrafish embryos. In rainbow trout which have received intraperitoneal injection of carbon particles, phagocytic cells were mainly present in the integument, including the skin and gut and particularly the gills, 4 days post-hatching. As the fish ages (18 days to 8 months), the kidney and spleen become the major sites of phagocytic cells (Zapata et al., 2006).

Some authors use the designation "kidney marrow" as the primary site for hematopoiesis in fish (versus the bone marrow in humans). The kidney is located in the body cavity in close contact with the osseous vertebral spine. An anterior segment contains predominantly hematopoietic tissue, and a middle and posterior segment are dominated by renal tissue but also contain hematopoietic tissue. Whereas the development of B cells occurs in the kidney, the mature T cells are generated in the thymus (Hordvik, 2009).

Other important organs in the immune system of fish are the spleen and mucosal tissues (gills, skin and gut). In the spleen, kidney and liver, aggregations of pigmented cells are present—called melanomacrophage centers. A lymphatic system sharing many morphological characteristics of the lymphatic vessels in other vertebrates is present in fish (Olson, 1996; Yaniv et al., 2006), but fish lack equivalents of the lymph nodes in mammals. However, lymphoid aggregates (consisting mainly of T cells) were recently detected in the gills of salmon (Haugarvoll et al., 2006). Lymphocytes differentiate in lymphoid organs at different times relative to hatching. In freshwater fish the thymus is the first organ to become lymphoid, followed by the blood and head kidney and after some delay in the spleen. In marine fish species the order in which the major lymphoid organs develop is kidney, spleen and finally, thymus (Zapata et al., 2006).

The specific antibody titres induced by immunization appear to increase with age of the fish, and the duration of protection increases with age. Very early exposure to antigens can induce tolerance. Size might correlate better than age since development is influenced by parameters such as temperature (Zapata et al., 2006). The onset of specific antibody production is dependent on the nature of the antigen (T cell dependent versus T cell independent) (Hordvik, 2009).

References

Aceto, A., F. Amicarelli, P. Sacchetta, B. Dragani, T. Bucciarelli, L. Masciocco, M. Miranda and C. Di Ilio. 1994. Developmental aspects of detoxifying enzymes in fish (*Salmo iridaeus*). Free. Rad. Res. 21(5): 285–294.

Ahmad, R.G. 2005. Is there a balance between oxidative stress and antioxidant defense system during development? Medical J. of Islamic World Academy of Sciences. 1(2): 55–63.

Andersson, T. and L. Förlin. 1992. Regulation of the cytochrome P450 enzyme system in fish. Aquat. Toxicol. 24(1-2): 1–19.

Andreasen, E.A., J.M. Spitsbergen, R.L. Tanguay, J.J. Stegeman, W. Heideman and R.P. Peterson. 2002. Tissue-specific expression of AHR2, ARNT2, and CYP1A in zebrafish embryos and larvae: Effects of developmental stage and 2,3,7,8-tetrachlorodibenzo-p-dioxin exposure. Toxicol. Sci. 68: 403–419.

Arukwe, A. 2001. Cellular and molecular responses to endocrine-modulators and the impact on fish reproduction. Mar. Pollut. Bull. 42(8): 643–655.

Arukwe, A. and A. Goksoyr. 2003. Eggshell and egg yolk proteins in fish: hepatic proteins for the next generation: oogenetic, population, and evolutionary implication of endocrine disruption. Comp. Hepatology. 2(4): 21 pp. Available at http://ww. Comparative-hepatology.com.

Arufe, M.I., J.M. Arellano, L. García, G. Albendín and C. Sarasquete. 2007. Cholinesterase activity in gilthead seabream (*Sparus aurata*) larvae: Characterization and sensitivity to the organophosphate azinphosmethyl. Aquat Toxicol. 84(3): 328–36.

Baker, S.M. and R. Mann. 1991. Metabolic rates of metamorphosis oysters (*Crassostrea virginica*) determined by microcalorimetry. Am. Zool. 31: 134A.

Belanger, S.E., E.K. Balon and J.M. Rawlings. 2010. Saltatory ontogeny of fishes and sensitive early life stages for ecotoxicology tests. 97(2): 88–95.

Bigsby, R., R.E. Chapin, G.P. Daston, B.J.Davis, J. Gorski, L.E. Gray, K.L. Howdeshell, R.T. Zoeller and F.S. vom Saal. 1999. Evaluation the effects of endocrine disruptors on endocrine function during development. Environ. Health Perspectives. 107(4): 613–618.

Bishop, R.E. and J.J. Torres. 1999. Leptocephalus energetics: metabolism and excretion. J. Experimental Biology. 202: 2485–2493.

Buckley, L.J. 1982. Effects of temperature on growth and biochemical composition of larval winter flounder *Pseudopleuronectes americanus*. Mar. Ecol. Prog. Ser. 8: 181–186.

Canepa, E., S. Fraschetti, S. Geraci, M. Licciano, M. Manganelli, G. Alberyelli and G. Riadi 1997. Microcalorimetry of some invertebrates: preliminary characterization of their metabolic activity during different developmental stages. Biol. Mar. Mediterr. 4: 626–628.

Cao, L., W. Huang, J. Liu, X. Yin and Sh. Dou. 2010. Accumulation and oxidative stress biomarkers in Japaneese flounder larvae and juveniles under cadmium exposure. Comp. Biochem. Physiol. 151C: 386–392.

Cazenave, J., M.A. Bistoni, E. Zwirnmann, D.A. Wunderlin and C. Wiegand. 2006. Attenuating effects of natural organic matter on microcystin toxicity in zebra fish (*Danio rerio*) embryos—benefits and costs of microcystin detoxication. Environmental Toxicology. 21(1): 22–32.

Clarke, M.E., C. Calvi, M. Domeier, M. Edmonds and P.J. Walsh. 1992. Effects of nutrition and temperature on metabolic enzyme activities in larval and juvenile red drum, *Sciaenops ocellatus*, and lane snapper, *Lutjanus synagris* Marine Biology. 112(1): 31–36.

Coelho, S., R. Oliveira, S. Pereira, C. Musso, I. Dominguesa, R.C Bhujelb, A.M. Soares and A.J.A. Nogueira. 2011. Assessing lethal and sub-lethal effects of trichlorfon on different trophic levels. Aquatic Toxicol. 103: 191–198.

Dabrowski, K. and A. Ciereszko. 2001. Ascorbic acid and reproduction in fish: endocrine regulation and gamete quality. Aquaculture Res. 32(8): 623–638.

Deane, E.E. and N.Y.S. Woo. 2003. Ontogeny of thyroid hormones, cortisol, hsp70 and hsp90 during silver sea bream larval development. Life Sci. 72: 805–818.

Desvilettes, C., G. Bourdier and J.C. Breton. 1997. Changes in lipid class and fatty acid composition during development in pike (*Esox lucius* L.) eggs and larvae. Fish Physiol. Biochem. 16: 381–393.

Domingues, I., R. Oliveira, J. Lourenço, C.K. Grisolia, S. Mendo and A.M. Soares. 2010. Biomarkers as a tool to assess effects of chromium (VI): comparison of responses in zebrafish early life stages and adults. Comp. Physiol. Biochem. 152C(3): 338–45.

Evans, R.P., C.C. Parrish, P. Zhu, J.A. Brown and P.J. Davis. 1997. Changes in phospholipase A activity and lipid content during early development of Atlantic halibut (*Hippoglossus hippoglossus*). Mar. Biol. 130: 369–376.

Falone, S., F. Cattani, M.T. Alamanou, A. Bonfigli, O. Zarivi, M. Mirandsa, A.M. Ragnelli and C. Di Ilio. 2004. Amphibian transition to the oxidant terrestrial environment affects the expression of glutathione S-transferases isoenzymatic pattern. Biochem. Biophys. Acta. 1691: 181–192.

Fiedler, T.J., M.E. Clarke and P.J. Walsh. 1998. Condition of laboratory-reared and wild-caugh larval Atlantic menhaden *Brevoortia tyrannus* as indicated by metabolic enzyme activities. Mar. Ecol. Prog Ser. 175: 51–66.

Finn, R.N. 2007. The physiology and toxicology of salmonid eggs and larvae in relation to water quality criteria. Aquatic Toxicol. 81: 337–354.

Finn, R.N., H.J. Fyhn, R.J. Henderson and M.S. Evjen. 1996. The sequence of catabolic substrate oxidation and enthalpy balance of developing embryos and yolk sac larvae of turbot (*Scophthalmus maximus* L). Comp. Biochem. Physiol. 115A(2): 133–151.

Fleming, C.R. and R.T. Di Giulio. 2011. The role of CYP1A inhibition in the embryotoxic interactions between hypoxia and polycyclic aromatic hydrocarbons (PAHs) and PAH mixtures in zebrafish (*Danio rerio*). Ecotoxicology. 20: 1300–1314.

Fluck, R.A. 1982. Localization of acetylcholinesterase activity in young embryos of the medaka *Oryzias latipes*, a teleost. Comp. Biochem. Physiol. 72C: 59–64.

Geubtner, J.A. 1998. Specific dynamics action, growth and development in larvae Atlantic cod *Gadus morhua*. Thesis B.A. University Delaware. 66 pp.

Goksøyr, A. 2006. Endocrine disruptors in the marine environment: mechanisms of toxicity and their influence on reproductive process in fish. J. Toxicol. Environ. Health. 69A: 175–184.

Goksøyr, A., T.S. Solberg and B. Serigstad. 1991. Immunochemical detection of cytochrome P450IA1 induction in cod larvae and juveniles exposed to a water soluble fraction of North Sea crude oil. Marine Pollution Bulletin. 22(3): 122–127.

Guerriero, G., R. Ferro, G.L. Russo and G. Ciarcia. 2004. Vitamin E in early stages of sea bass (*Dicentrarchus labrax*) development. Comp. Biochem. Physiol. 138A (4): 435–439.

Hamm, J.T., B.W. Wilson and D.E. Hinton. 2001. Increasing uptake and bioactivation with development positively modulate diazinon toxicity in early life stage medaka (*Oryzias latipes*). Toxicol. Sci. 61: 304–313.

Haugarvoll, E., J. Thorsen, M. Laane, Q. Huang and E.O. Koppang. 2006. Melanogenesis and evidence for melanosome transport to the plasma membrane in a CD83 teleost leukocyte cell line. Pigment Cell Res. 19: 214–25.

Hodgson, P.A. and S.G. George. 1998. Xenobiotic biotransformation enzyme gene expression in early larval stages of plaice. Mar. Environ. Res. 46(1-3): 465–468.

Hordvik, I. Development of the immune system in fish. 2009. In: The fish larvae: a transitional life form, the foundation for aquaculture and fisheries. Report from a working group on research on early life stages of fish. The Research Council of Norway. 64–66.

Hornung, M.W., P.M. Cook, P.N. Fitzsimmons, D.W. Kuehl and J.W. Nichols. 2007. Tissue distribution and metabolism of benzo[a]pyrene in embryonic and larval medaka (*Oryzias latipes*). Toxicol. Sci. 100(2): 393–405.

Hsu, H.-J., P. Hsiao, M.-W. Kuo and B.C. Chung. 2002. Expression of zebrafish cyp11a1 as a maternal transcript and in yolk syncytial layer. Gene Expr. Patterns. 2: 219–222.

Hu, D., E. Klann and E. Thiels. 2007. Superoxide dismutase and hippocampal function: age and isozyme matter. Antioxidants & Redox Signaling. 9: 201–210.

Huang, W., L. Cao, J. Liu, L. Lin and S. Dou. 2010. Short-term mercury exposure affecting the development and antioxidant biomarkers of Japanese flounder embryos and larvae. Ecotoxicology and Environmental Safety. 73: 1875–1883.

Incardona, J.P., M.G. Carls, H. Teraoka, C.A. Sloan, T.K. Collier and N.L. Scholz. 2005. Aryl hydrocarbon receptor-independent toxicity of weathered crude oil during fish development. Environ. Health Perspectives. 1131(2): 1755–1762.

Isuev, A.R., L.V. Ponomareva, G.V. Kossova, S.V. Kotelevtzev and A.M. Beim. 1991. Membrane lipids and monooxygenase of *Misgurnus fossilis* in normal conditions and after injection of 3-metylcholarntren and sovol. Biological Sciences. 1: 32–37 (*in Russian*).

Jonsson, M.E., M.J. Jenny, B.R. Woodin, M.E. Hahn and J.J. Stegeman. 2007. Role of AHR2 in the expression of novel cytochrome P450 1 family genes, cell cycle genes, and morphological defects in developing zebra fish exposed to 3,3′,4,4′,5-pentachlorobiphenyl or 2,3,7,8-tetrachlorodibenzo-p-dioxin. Toxicol. Sci. 100(1): 180–193.

Kadomura, K., T. Nakashima, M. Kurachi, K. Yamaguci and T. Oda. 2006. Production of reactive oxygen species (ROS) by devil stinger (*Inimicus japonicus*) during embryogenesis. Fish and Shellfish Immunology. 21(2): 209–214.

Kalaimani, N., N. Chakravarthy, R. Shanmugham, A.R. Thirunavukkarasu, S.V. Alavandi and T.S. Santiago. 2008. Anti-oxidant status in embryonic, post-hatch and larval stages of Asian sea bass (*Lates calcarifer*). Fish Physiol. Biochem. 34: 151–158.

Kalinina, E.M. Reproduction and development of Azov and Black Sea gobies. 1976. Kiev, Naukova Dumka Publ. 120 pp. (*in Russian*).

Kashiwada, S., K. Goka, H. Shiraishi, K. Arizono, K. Ozato, Y. Wakamatsu and D.E. Hinton. 2007. Age-dependent *in situ* hepatic and gill CYP1A activity in the see-through medaka (*Oryziasla*). Comp. Biochem. Physiol. 145C(1): 96–102.

Kishida, M. and G.V. Callard. 2001. Distinct cytochrome P450 aromatase isoforms in zebrafish (*Danio rerio*) brain and ovary are differentially programmed and estrogen regulated during early development. Endocrinology. 142(2): 740–750.

Kitano, T., K. Takamun, T. Kobayashi, Y. Nagahama and S.I. Abe. 1999. Supression of P450 aromatase gene expression in sex-reversed males produced by rearing genetically female larvae at a high water temperature during a period of sex differentiation in the Japanese flounder (*Paralichthys olivaceus*). J. Molecular Endocrinology. 23: 167–176.

Konovalov, Ju.D. 1984. SH- and SS-groups in proteins and low molecular weight non-protein thiol components in early development of fish. Advances of Biology. 98. 2(5): 267–282 (*in Russian*).

Koponen, K., P. Lindstrom-Seppaa and J.V.K. Kukkonen. 2000. Accumulation pattern and biotransformation enzyme induction in rainbow trout embryos exposed to sublethal aqueous concentrations of 3; 30,4; 40-tetrachlorobiphenyl. Chemosphere. 40: 245–253.

Krautz, M.C., S. Vasquez, L.R. Castro, M. Gonzalez, A. Llanos-Rivera and S. Pantoja. 2010. Changes in metabolic substrates during early development in anchoveta *Engraulis ringens* (Jenyns 1842) in the Humboldt Current. Mar. Biol. 157: 1137–1149.

Kuzminova, N.S. 2003. The influence of different concentration of sewage on marine organisms. Abstracts of the Int. Conf. "Space and Biosphere" Partenit, Ukraine. Sept. 28–Oct. 4. Simpheropol. 164.

Lamprecht, I. 1998. Monitoring metabolic activities of small animals by means of microcalorimetry. Pure & Appl. Chem. 70(3): 695–700.

Laurel, B.J., L.A. Copeman, Th.P. Hurst and Ch.C. Parrish. 2010. The ecological significance of lipid/fatty acid synthesis in developing eggs and newly hatched larvae of Pacific cod (*Gadus macrocephalus*). Mar. Biol. 157: 1713–1724.

Lemieux, H., N.R.L. François and P.U. Blier. 2003. The early ontogeny of digestive and metabolic enzyme activities in two commercial strains of arctic charr (*Salvelinus alpinus* L.). J. Exp. Zool. 299A: 151–160.

Lin, H.C., S.C. Hsu and P.P. Hwong. 2000. Maternal transfer of cadmium tolerance in larval *Oreochromis mossambicus*. Journal of Fish Biology. 57: 239–249.

Luckenbach, T.H. Ferling, E.M. Gernhofer, H.-R. Koehler, R.-D. Negele, E. Pfefferle and R. Triebskorn. 2003. Developmental and subcellular effects of chronic exposure to sub-lethal concentrations of ammonia, PAH and PCP mixtures in brown trout (*Salmo trutta f. fario* L.) early life stages. Aquat. Toxicol. 65: 39–54.

McCollum, A., J. Geubtner and H. von Herbing. 2006. Metabolic cost of feeding in Atlantic Cod (*Gadus morhua*) larvae using microcalorimetry. ICESJ. Marine Science. 63(3): 333–339.

Mekkawy, I.A.A. and F. E.-D. Lashein. 2003. The effect of lead and cadmium on LDH and G-6-PDH isoenzyme patterns exhibited during the early embryonic development of the teleost fish, *Ctenopharyngodon idellus* with emphasis on the corresponding morphological variations. In: The Big Fish Bang. Proceedings of the 26th Annual Larval Fish Conference. 2003. Edited by Howard I. Browman and Anne Berit Skiftesvik Published by the Institute of Marine Research, Postboks 1870 Nordnes, N-5817, Bergen, Norway. 275–292.

Mommsen, T.P. and P.J. Walsh. 1988. Vitellogenesis and oocyte assembly. In: Fish Physiology: The physiology of developing fish. Part A. Eggs and larvae, Vol. XI (W.S. Hoar and D.J. Randall eds.). pp. 347–406. San Diego: Academic Press.

Monod, G., M.-A. Boudry and C. Gillet. 1996. Biotransformation enzymes and their induction by β-naphthoflavone during embryo-larval development in salmonid species. Comp. Biochem. Physiol. 114C(1): 45–50.

Moskalkova, K.I. 1984. Unusual way of embryonic feeding in *Neogobius melanostomus* (Pallas) (Pisces, Gobiidae). Reports of the USSR Academy of Science. 278: 1127–1130 (*in Russian*).

Moskalkova, K.I. 1985. "Couprofagia" in embryos of teleost fish *Neogobius melanostomus* (Pallas) (Pisces, Gobiidae). Reports of the USSR Academy of Sciences. 282: 1251–1254 (*in Russian*).

Mourente, G. and R. Vazquez. 1996. Changes in the content of total lipid, lipid classes and fatty acids of developing eggs and unfed larvae of the Senegal sole, *Solea senegalensis* Kaup. Fish Physiol. Biochem. 15: 221–235.

Mourente, G., A. Rodriguez, A. Grau and E. Pastor. 1999a. Utilization of lipids by *Dentex dentex* L. (Osteichyhyes, Sparidae) larvae during lecitotrophia and subsequent starvation. Fish Physiol. Biochem. 21: 45–58.

Mourente, G., D.R. Tocher, E. Diaz, A. Grau and E. Pastor. 1999b. Relationships between antioxidants, antioxidant enzyme activities and lipid peroxidation products during early development in *Dentex dentex* eggs and larvae. Aquaculture. 179(1-4): 309–324.

Mourente, G., E. Diaz-Salvago, D.R. Tocher and J.G. Bell. 2000. Effects of dietary polyunsaturated fatty acid/vitamin E (PUFA/tocopherol ratio on antioxidant defense mechanisms of juvenile gilthead sea bream (*Sparus aurata* L., Osteichthys, Sparidae). Fish Physiol. Biochem. 23: 337–351.

Nakano, E. and A.H. Whiteley. 1965. Differentiation of multiple molecular forms of four dehydrogenases in the teleost, *Oryzias latipes*, studied by disc electrophoresis. J. Exp. Zool. 159: 167–179.

Nam, B.H., I. Hironi and T. Aoki. 2003. The four TCR genes of teleost fish: the cDNA and genomic DNA analysis of Japanese flounder (*Paralichthys olivaceus*) TCR alpha-,beta-, gamma-, and delta-chains. J. Immunol. 170: 3081–90.

Naz, M. 2009. Ontogeny of biochemical phases of fertilized eggs and yolk sac larvae of gilthead seabream (*Sparus aurata* L.). Turkish Journal of Fisheries and Aquatic Sciences. 9: 77–83.

Oliveira, R., I. Domingues, C.K. Grisolia and A.M. Soares. 2009. Effects of triclosan on zebrafish early-life stages and adults. Environ. Sci. Pollut. Res. 16: 679–688.

Olson, K.R. 1996. Secondary circulation in fish: anatomical organization and physiological significance. J. Exp. Zool. 275: 172–185.

Olsson, P.E., M. Zafarullah, R. Foster, T. Hamor and L. Gedamu. 1990. Developmental regulation of metallothionein mRNA, zinc and copper levels in rainbow trout, *Salmo gairdneri*. European Journal of Biochemistry. 193: 229–235.

Ornsrud, R., A. Wargelius, O. Sale, K. Pittman and R. Waagbo. 2004. Influence of egg vitamin A status and egg incubation temperature on subsequent development of the early vertebral column in Atlantic salmon fry. J. Fish biology. 64(2): 399–417.

Ortiz-Delgado, J.B. and C. Sarasquete. 2004. Toxicity, histopathological alterations and immunohistochemical CYP1A induction in the early life stages of the sea bream, *Sparus aurata*, following waterborne exposure to B(a)P and TCDD. J. Mol. Histol. 35: 29–45.

Otto, D.M.E. and T.W. Moon. 1996. Endogenous antioxidant system of two teleost fish, the rainbow trout and the black bullhead, and the effect of age. Fish Physiol. Biochem. 15 (4): 349–358.

Overnell, J.R.S. Batty. 2000. Scaling of enzyme activity in larval herring and plaice: effects of temperature and individual growth rate on aerobic and anaerobic capacity. Journal of Fish Biology. 56: 577–589.

Ozernuk, N.D. 1985. Energetic metabolism of fish in early ontogenesis. Nauka. Moscow. 175 pp (*in Russian*).

Pakkasma, S., O.-P. Penttinen and J. Piironen. 2006. Metabolic rate of Arctic charr eggs depend on their parentage. J. Comp. Physiol. 176B: 387–391.

Palikova, M., R. Krejai, K. Hilscherova, B. Buryskova, P. Babica, S. Navratil, R. Kopp and L. Blaha. 2007a. Effects of different oxygen saturation on activity of complex biomass and aueous crude extract of cyanobacteria during embryonal development in carp (*Cyprinus carpio* L.) Acta Vet. Brno. 76: 291–299.

Palikova, M., R. Krejcı, K. Hilscherova, P. Babica, S. Navratil, R. Kopp and L. Blaha. 2007b. Effect of different cyanobacterial biomasses and their fractions with variable microcystin content on embryonic development of carp (*Cyprinus carpio* L.). Aquatic Toxicol. 81: 312–318.

Pauka, L.M., M. Maceno, S.C. Rossi and C. Silva de Assis. 2011. Embryotoxicity and biotransformation responses in zebrafish exposed to water-soluble fraction of crude oil. Bull. Environ. Contam. Toxicol. 86: 389–393.

Penttinen, O.-P. and J. Holopainen. 1995. Physiological energetics of a midge, *Chironomus riparius* Meigen (Insecta, Diptera): normoxic heat output over the whole life cycle and response of larva to hypoxia and anoxia. Oecologia. 103: 419–424.

Peters, L.D., C. Porte, J. Albaiges and D.R. Livingstone. 1994. 7-etoxyresorufin O-deethylase (EROD) and antioxidant enzyme activities in larvae of sardine (*Sardina pilchardus*) from the North Coast of Spain. Mar. Pollut. Bull. 28(5): 299–304.

Peters, L.D. and D.R. Livingstone. 1996. Antioxidant enzyme activities in embryologic and early larval stages of turbot. J. Fish Biol. 49: 986–997.

Peters, L.D., C. Porte and D.R. Livingstone. 2001. Variation of antioxidant enzyme activities of sprat *Sprattus sprattus* larvae and organic contaminant levels in mixed zooplankton from the Southern North Sea. Mar. Pollut. Bull. 42(11): 1087–1095.

Pethybridge, H., R. Daley, P. Virtue and P.D. Nichols. 2011. Lipid (energy) reserves, utilization and provisioning during oocyte maturation and early embryonic development of deepwater chondrichthyans. Mar. Biol. 158: 2741–2754.

Pfeiler, E. 1986. Towards an explanation of the developmental strategy in leptocephalous larvae of marine teleost fishes. Envir. Biol. Fish. 15: 3–13.

Pfeiler, E. 1996. Energetics of metamorphosis in bonefish (*Albula* sp.) leptocephali: Role of keratan sulfate glycosaminoglycan. Fish Physiol. Biochem. 15: 359–362.

Phillips, T.A., R.C. Summerfelt and G.J. Atchison. 2002a. Environmental, biological, and methodological factors affecting cholinesterase activity in walleye (*Stizostedion vitreum*) Archiv. Environ. Contam. Toxicol. 43(1). 75–80.

Phillips, T.A., J. Wu, R.C. Summerfelt and G.J. Atchinson. 2002b. Acute toxicity and cholinesterase inhibition in larval and early juvenile walleye exposed to chlorpyrifos. Environ. Toxicol. Chem. 21(7): 1469–1474.

Powell, W.H., R. Bright, S.M. Bello and M.E. Hahn. 2000. Developmental and tissue-specific expression of AHR1, AHR2, and ARNT2 in dioxin-sensitive and -resistant population of the marine fish *Fundulus heteroclitus*. Toxicol. Sci. 57: 229–239.

Riggio, M., S. Filosa, E. Parisi and R. Scudiero. 2003. Changes in zinc, copper and metallothionein contents during oocyte growth and early development of the teleost *Danio rerio* (zebrafish). Comp. Biochem. Physiol. 135C: 191–196.

Roche-Mayzaud, O., P. Mayzaud and C. Audet. 1998. Changes in lipid classes and trypsin activity during the early development of brook charr, *Salvelinus fontinalis* (Mitchill), fry. Aquacult. Res. 29: 137–152.

Rønnestad, I. and H.J. Fyhn. 1993. Metabolic aspects of free amino acids in developing marine fish eggs and larvae. Rev. Fish Sci. 1(3): 239–259.

Ronnestad, I., K. Hamre, O. Lie and R. Waagbo. 2005. Ascorbic acid and α-tocopherol levels in larvae of Atlantic halibut before and after exogenous feeding. J. Fish Biol. 55(4): 720–731.

Rudneva, I.I. 1994. The ratio of antioxidant enzyme activities and lipid peroxidation in developing eggs of Black Sea round goby. Ontogenesis. 25: 13–20 (*in Russian*).

Rudneva, I.I. 1995. The dynamics of antioxidant enzyme activities at early development of several Black Sea fish species. Ukrainian Biochem. J. 67(1): 12–15 (*in Russian*).

Rudneva, I.I. 1997. Development of antioxidant enzyme system in early life of marine animals. Achievements in Biology. 117(3): 390–398 (*in Russian*).

Rudneva, I.I. 1999. Antioxidant system of Black Sea animals in early development. Comp. Biochem. Physiol. C. 122: 265–271.

Rudneva, I.I. and V.G. Shaida. 2000. Microcalorimetric studies of marine organisms early development. Microcalorimetric Studies in Marine Biology. ECOSY. Sevastopol. 139–149 (*in Russian*).

Rudneva, I.I., T.L. Chesalina and V.G. Shaida. 2001. Physiological and biochemical characteristics of Black Sea flounder *Psetta maxima maeotica* embryogenesis. Ichthyology. 41: 717–720 *(in Russian)*.

Rudneva, I.I., V.G. Shaida and N.S. Kuz'minova. 2004. The fungicide cuprocsat effect on the Artemia and Atherina larvae heat production. Agroecological J. 3: 81–83 *(in Russian)*.

Rudneva, I.I. and V.G. Shaida. 2006. Metabolic strategy in the early life of some Black Sea fish species measured by microcalorimetry method. XIVth Conference The Amber ISBC. Abstr. Sopot, Poland, June 2–6, 2006. P.74.

Rudneva, I.I. and V.G. Shaida. 2012. Metabolic rate of marine fish in early life and its relationship to their ecological status. In: Fish Ecology (Ed. Gempsy P.). Nova Science Publishers. New York, pp. 1–29.

Russel, M., J. Yao, H. Chen, F. Wang, Y. Zhou, M.M. Choi, G. Zaray and P. Trebse. 2009. Different technique of microcalorimetry and their application to environmental sciences: a review. J. American Science. 5(4): 194–208.

Samsonova, M.V., T.I. Lapteva and Yu.B. Filippovich. 2005. Aminotransferases in early development of Salmonid fish. Ontogenez. 36(2): 96–101.

Scheil, V., C. Keinle, R. Osterauer, A. Gerhardt and H.-R. Kohler. 2009. Effects of 3,4-dichloroaniline and diazinon on different biological organization levels of zebrafish *(Danio rerio)* embryos and larvae. Ecotoxicology. 18: 355–363.

Scholz, S., S. Fischer., U. Gundel, E. Kuster, T. Luckenbach and D. Voelker. 2008. The zebrafish embryo model in environmental risk assessment—applications beyond acute toxicity testing. Environ. Sci. Pollut. Res. 15: 394–404.

Schmolz, E., S. Drutschmann, B. Schricker and L. Lamprecht. 1999. Calorimetric measurements of energy contents and heat production rates during development of the wax moth *Galleria mellonella*. Thermochimica Acta. 337: 83–88.

Selman, K. and R. Wallace. 1989. Cellular aspects of oocyte growth in teleosts. Zoological Science. 6: 211–231.

Sole, M., J. Portrykus, C. Fernandez-Diaz and J. Blasco. 2004. Variations on stress defenses and metallothionein levels in the Senegal sole, *Solea senegalensis*, during early larval stages. Fish Physiol. Biochem. 30: 57–66.

Sullivan, C. and C.H. Kim. 2008. Zebrafish as a model for infectious disease and immune function. Fish Shellfish Immunol. 25: 341–50.

Takahashi, A., H. Totsuni-Nakano, M. Nakano, S. Mashiko, N. Suzuki, C. Ohma and H. Inaba. 1989. Generation of 0_2 and tyrosine cation-mediated chemiluminescence during the fertilization of sea urchin eggs. FEBS Lett. 246(1-2): 117–119.

Terova, G., M. Saroglia, Z.G. Papp and S. Cecchini. 1998. Ascorbate dynamics in embryos and larvae of sea bass and sea bream, originating from broodstocks fed supplements of ascorbic acid. Aquaculture International. 6: 357–367.

Tocher, D.R. 2003. Metabolism and function of lipids and fatty acids in teleost fish. Rev. Fish Sci. 11: 107–184.

Trant, J.M., S. Gavasso, J. Ackers, B.-Ch. Chung and A.R. Place. 2001. Development expression of cytochrome P450 aromatase genes (CYP19a and CYP19b) in zebrafish fry *(Danio rerio)*. J. Exp. Zool. 290: 475–483.

Tyler, Ch. R., R. van Aerle, T.H. Hutchinson, S. Maddix and H. Trip. 1999. An vivo testing system for endocrine disruptors in fish early life stages using induction of vitellogenin. Environ. Toxicol. Chem. 18(2): 337–347.

Uesugi, S. and S. Yamazoe. 1964. Acetylcholinesterase activity during the development of the rainbow trout eggs. Gunma J. Med. Sci. 13: 91–93.

Vifia, J., F. Pallardo and C. Borras. 2003. Mitochondrial theory of aging: importance to explain why females live longer than males. Antioxidnts & Redox Signaling. 5: 549–556.

Wegwood, S., S. Lakshminrusimha, T. Fukai, J. Russell, P. Schumacker and R. Steinhorn. 2011. Hydrogen peroxide regulates extracellular superoxide dismutase activity and expression in neonatal pulmonary hypertension. Antioxidants and Redox Signaling. 15(6): 1497–1506.

Wiegand, C., S. Pflugmacher, A. Oberemm, N. Meems, K.A. Beattie, C.E.W. Steinberg and A.G. Codd. 1999. Uptake and effects of microcystin-LR on detoxication enzymes of early life stages of the zebra fish (*Danio rerio*). Environ. Toxicol. 14(1): 89–95.

Wiegand, C., S. Pflugmacher, A. Oberemm and Ch. Steiberg. 2000. Activity development of selected detoxification enzymes during the ontogenesis of the Zebrafish (*Danio rerio*). Int. Review of Hydrobiology. 85: 413–422.

Werner, J., K. Wautier, R.E. Evans, C.L. Baron, K. Kidd and V. Palace. 2003. Waterborne ethynylestradiol induces vitellogenin and alters metallothionein expression in lake trout (*Salvelinus namaycush*). Aquat. Toxicol. 62: 321–328.

Whyte, J., W. Clarke, N. Ginther and J. Jensen. 1993. Biochemical changes during embryogenesis of the Pacific halibut, *Hippoglossus stenolepis* (Schmidt). Aquaculture Research. 24: 193–201.

Winston, G.W. Oxidants and antioxidants in aquatic organisms. 1991. Comp. Biochem. Physiol. 100C (1-2): 173–176.

Wu, S.M., Y.-C. Ho and M.J. Shih. 2007. Effects of Ca^{2+} or Na^+ on metallothionein expression in tilapia larvae (*Oreochromis mossambicus*) exposed to cadmium or copper. Arch. Environ. Contam. Toxicol. 52: 229–234.

Wu, Y.D., L. Jiang, Z. Zhou, M.H. Zheng, J. Zhang and Y. Liang. 2008a. CYP1A/regucalcin gene expression and edema formation in early stages of *Oreochromis mossambicus*. Bull. Environ. Contam. Toxicol. 80: 482–486.

Wu, S.M., H.C. Lin and W.L. Yang. 2008b. The effects of maternal Cd on the metallothionein expression in tilapia (*Oreochromis mossambicus*) embryos and larvae. Aquatic Toxicology. 87: 296–302.

Yanity, K., S. Isogai, D. Castranova, L. Dye, J. Hitomi and B.M. Weinstein. 2006. Life imaging of lymphatic development in the zebrafish. Nat. Med. 12: 711–716.

Yun, L., L. Moore, E.O. Koppang and I. Hordvick. 2008. Characterization of the CD3zeta, CD3gamma delta and CD3epsilon subunits of the T cell receptor complex in Atlantic salmon. Dev. Comp. Immunol. 32: 26–35.

Zapata, A., B. Diez, T. Cejalvo, C. Gutierrez-de Frias and A. Cortes. 2006. Ontogeny of the immune system of fish. Fish Shellfish Immunol. 20: 126–36.

Zhu, P., R.P. Evans, C.C. Parrish, J.A. Brown and P.J. Davis. 1997. Is there a direct connection between amino acid and lipid metabolism in marine fish embryos and larvae? Bull. Aquacult. Assoc. Can. 97-2. 48–50.

Zlabek, V., T. Randak., J. Kolarova, Z. Svobodova and H. Kroupova. 2009. Sex differentiation and vitellogenin and 11-ketotestosterone levels in chub, *Leuciscus cephalus* L., exposed to 17 β-estradiol and testosterone during early development. Bull. Environ. Contam. Toxicol. 82: 280–284.

CHAPTER 4

Biomarkers Responses in Fish Embryos and Larvae

4.1 Experimental Studies

The sensitivity of early development stages to various pollutants has been frequently reported for many aquatic animals such as sea urchins, bivalves, corals and fish species (Bellas et al., 2005; Farina et al., 2008; Cao et al., 2010; Wright, 1995). Larvae have important roles in the life cycle of marine organisms as during this phase there are dramatic morphological and physiological changes that lead to the growth of adult form. Despite their importance, data concerning the toxicity of pollutants in larvae is scarce (Bellas et al., 2005; Farina et al., 2008). During this phase the organisms are less tolerant than embryos or adults to the various kinds of toxicants. The resistance of fish embryos and larvae to stressors depends on their taxonomic and phylogenic position. For instance, the cartilaginous fishes are characterized by life cycle attributes that result in low intrinsic rates of increase, making their population highly sensitive to environmental impacts. These qualities include low fecundity that is expressed in the production of a relatively small number of large, heavily yolked eggs in the oviparous species, which comprise about 40% of the sharks and all skates. These embryos develop slowly following their laying in nursery grounds, commonly found in shallow coastal waters that are highly impacted due to human activity. The distribution of *Chondrichthyan* eggs in polluted habitats could enhance their exposure to elevated contaminant levels during this long early phase of development which is most susceptible to environmental pollutants in teleost fish, particularly for metals (Jeffree et al., 2008). Structural and functional characteristics of fish eggs demonstrated that the following attributes may contribute to high accumulation of pollutants (Knight et al., 1996; Jeffree et al., 2008):

- the indicated oxidative phenolic cross-linking of the case, in combination with bound catechol groups in the outermost layer, to facilitate the chelation of metals from seawater;
- the presence of disulphide bonds that act in the stabilization of egg-case collagen could be expected to provide strong binding sites for metalchelation, particularly of inorganic Hg and Pb;
- the high internal surface area that is created by the open meshwork structure of the collagen fibres that make up the laminae of the egg-case layers would increase sites for the potential absorption of elements passing through the transverse channels in the wall.

Hence, there are many factors including exogenous and endogenous which influence fish sensitivity to environmental stressors.

4.1.1 Heavy metals effects

The heavy metals listed by the Environment Protection Agency (EPA) as the main pollutants of the environment are the following: As, Cd, Cr, Cu, Hg, Ni, Pb, and Zn. They are readily taken up by the eggs and larvae and impair the normal physiological functions of Ca^{2+} and Mg^{2+} by binding to their receptor sites. Metal ions have been shown to impair vertebral column formation as well as growth, metabolism, yolk resorption, or hydro-mineral and acid–base balance by inhibiting the activity of ion pumps in the gill or yolk-sac epithelia. The least sensitive phases appear to be the early eggs, where ambient metal concentrations apparently do not interfere with fertilization (Finn, 2007). There are several factors including abiotic (physical and chemical parameters of the water and sediments in marine location), biotic (specificity of fish biology, feeding behavior, age, sex, swimming activity and metabolic rate) and anthropogenic (level of pollution led human activity) that may affect trace elements accumulation in marine organisms (Sole et al., 2009). Opposite, low pH, humic acid, diethylene triamine pentaacetic acid (DTPA), malachite green, dissolved organic carbon, chloride concentration, and temperature have been shown to reduce the uptake of heavy metals by the eggs and larvae. The reduced uptake is associated with complexing of the metal ions with organic matter in the water, although these protective effects appear more important for acute rather than chronic exposures to dissolved metals (Finn, 2007).

Aquatic organisms, such as fish, accumulate pollutants directly from contaminated water and indirectly via the food chain. The most important factor is diet (Mathews and Fisher, 2009). In addition, radiotracing experiments demonstrated that eggs of some cartilaginous fish species (spotted dogfish *Scyliorhinus canicula*) absorbed Pb, inorganic Hg, Cs, Mn, Zn, Co directly from the seawater (Jeffree et al., 2008). It is well known that animals which include a large quantity of fish in their diet have the tendency to accumulate mercury to higher concentrations while higher cadmium level was found in fish in which diet invertebrates predominate (Storelli et al., 2005).

However, early life stages of fish demonstrate the resistance to toxic compounds including heavy metals. Part of the tolerance to metal intoxication may lie with maternal transfer of metallothionein genes as found in other teleosts. The binding properties of the chorion provide some protection for the embryo (Finn, 2007).

Fish have been recognized as the most important source of mercury in aquatic environments while invertebrates (mollusks and crustacea) are the important source for cadmium (Marcovecchio et al., 1991). A number of environment pollutants including heavy metals can cause oxidative stress in aquatic organisms (Fig. 4.1). Fish living in marine locations contaminated with different xenobiotics may be exposed to oxidative stressors caused by a variety of oxyradicals. To protect this chemical compounds aquatic organisms have developed different mechanisms such as the induction of

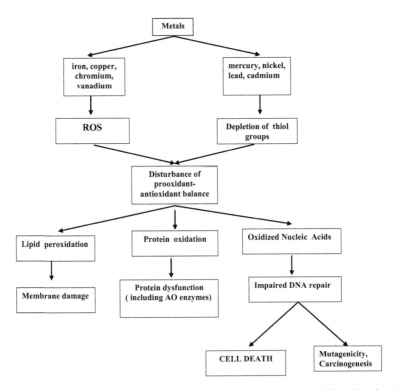

Fig. 4.1. Possible mechanisms for metal-induced oxidative stress (adopted from Ercal et al., 2001).

antioxidant enzymes including superoxide dismutase (SOD), catalase (CAT), peroxidase (PER) and glutathione (GSH) related enzymes (glutathione reductase (GR), glutathione peroxidase (GP) (Livingstone, 2001). Alterations in the antioxidant enzyme activities of aquatic animals in response to pollutants are used to indicate the potential for more severe hazards.

Among chemical elements mercury, lead, cadmium and arsenic are non-essential elements while Cu, Zn, Co, Cr, Fe, Mn, Se are essential. Many enzymes including SOD contain Zn, Mn and Cu. Because they are essential elements their concentrations can be regulated in the organism and thus their levels cannot be used for biomonitoring purposes (Barata et al., 2005) while the other trace elements can successfully apply to the evaluation of the marine environment. Fish embryos and larvae are the most sensitive to environmental pollutants, thus they have been widely used as bioindicators for water quality evaluation.

Metals involve in oxidative damage because they produce free radicals in two ways. Redox active metals such as iron, copper, chromium, and vanadium generate ROS through redox cycling. Metals without redox potential, such as mercury, nickel, lead, and cadmium, impair antioxidant defenses, especially those involving thiol-containing antioxidants and enzymes. An important mechanism of free radical production is the Fenton reaction, by which ferrous iron (II) is oxidized by hydrogen peroxide to ferric iron (III), a hydroxyl radical, and a hydroxyl anion. The superoxide radical can reduce iron to its ferrous form. Copper, chromium, vanadium, titanium, cobalt, and their complexes can also be involved in the Fenton reaction (Ercal et al., 2001; Lesser, 2006; Porte et al., 2000; Lushchak, 2011; Sevcikova et al., 2011). Activation of redox-sensitive transcription factors such as AP-1, p53, and NF-κB is another mechanism by which metals can participate in producing oxidative stress. These transcription factors control the expression of protective genes which repair DNA and influence apoptosis, cell differentiation, and cell growth (Valko et al., 2005). The main redox- active metals are the following iron, copper, zinc, manganese and chromium; while cadmium, mercury, lead, arsenic, selenium and some others are redox-inactive (Sevcikova et al., 2011). On the other hand, some metals are essential (Fe, Cu, Zn, Mn, Mg, Se, Co, Ni) while the others are non-essential and toxic (Hg, Cd, As, Pb). The toxic effects of heavy metals have been well documented, and many of these and other environmental pollutants are known to be embryotoxic or teratogenic. However, it is difficult to identify individual cells that respond to toxicants among the wide range of cell populations in an intact animal, particularly during early development when cells are continually change their molecular and physiological characteristics during differentiation process (Blechinger et al., 2002).

Essential metals

Iron

Iron is an essential element required for many physiological functions and it exists in many enzymes (hemoglobin, cytochromes, peroxidases, catalases). Its homeostasis is strictly regulated by various mechanisms. In biological system iron exists in three oxidation states (II, III, and IV). The majority of iron in the organism is bound to hemoglobin, transferrin, ferritin, and iron-containing enzymes. Therefore, only a trace amount of free iron is present. The deleterious effects of iron include DNA damage, lipid peroxidation (LPO), and oxidation of proteins via the Fenton reactions (Valko et al., 2005). Excessive uptake of iron or disturbances in its regulation can be toxic which is related to its ability to catalyze ROS formation via the Fenton reaction. Iron may also potentiate the toxicity of chemicals such as paraquat or 2,3,7,8-tetrachlorodibenzo-*p*-dioxin (Sevcikova et al., 2011).

Lipid peroxidation and modification in antioxidant enzyme activity in embryos of medaka *Oryzias latipes* exposed to nano-iron was observed (Li et al., 2009). Dose-dependent inhibition of SOD activity and increased production of TBA-reactive products (MDA) were indicated in medaka embryos. The researchers documented that medaka embryos are more sensitive to nano-iron exposure than the adults.

Copper

Copper plays an essential role in a variety of metabolic processes. It is a component of many enzymatic and structural proteins, including Cu-Zn SOD, cytochrome oxidase, and ceruloplasmin. The cellular toxicity of copper can be explained through its participation in the Fenton reaction. Cu (I) ion can catalyze the formation of hydroxyl radicals. Copper binds thiol-containing molecules such as glutathione. On the other hand, copper plays a protective role against oxidative damage caused by variety of xenobiotics. The antioxidant effect of ceruloplasmin and metallothioneins seems to be the mechanism by which copper protects under such conditions. Ceruloplasmin serves as a transport protein of copper in plasma and copper pre-exposure increases the activity of ceruloplasmin in adult fish serum (Sevcikova et al., 2011).

High concentration of copper at low pH demonstrated toxic effects in fish because it leads to the imbalance of redox cycle, respiration, cartilage formation, etc. Sublethal Cu concentrations disrupt ion balance and growth in larvae, juvenile, and adult fish of tilapia. Metallothioneins play a major role in the detoxification of copper. MT significantly increased by 3.3 to 4.3- and 6.0 to 7.7-fold compared to controls after exposure of tilapia larvae to 75 µg/L Cu^{2+} and 75 µg/L Cu^{2+} + 0.52 mM Na$^+$. Metallothionein expression was significantly higher in Cu/hyper-Na-treated tilapia larvae than in the group treated with Cu only (Wu et al., 2007). We also investigated the effect of Cu-containing pesticide cuprocsat (CuSO4 3Cu(OH)2 1/2H2O) on *Atherina* larvae and the heat production of treated fish was lower as compared to the impact larvae.

Chromium

The most biologically important oxidative states of chromium are trivalent (Cr III) and hexavalent (Cr VI). The trivalent and hexavalent forms of chromium are involved in redox cycle. Cell reducing agents such as GSH and nicotinamideadenine dinucleotidephosphate (NADPH) reduce Cr (VI) to the pentavalent state (V), which can participate in the Fenton reaction to produce hydroxyl radicals. The hexavalent form can be reduced to the trivalent form. This transformation is considered to be a major means of detoxification of Cr (VI) in biological systems. Chromium (III) plays a regulatory role in physiological glucose metabolism. Chromium (VI) actively enters cells through an anion (phosphate) transport mechanism. Chromium (III), meanwhile, is not able to use this mechanism (Valko et al., 2005). Many reports suggest that oxidative-induced alterations of DNA are the main effect of chromium in the studied adult fish species (Sevcikova et al., 2011). The information of chromium toxic effects and biomarker response in fish early life stages is very limited, but we could propose that the effects are similar as in adult animals.

Manganese

Manganese is a relatively common, yet poorly studied element in both freshwater and marine ecosystems, where it can be significantly bioconcentrated. The knowledge about the mechanisms of Mn toxicity on fish health and particularly in early developing stages is still limited. The manganese induced oxidative stress and the antioxidant response after a 96h waterborne Mn-exposure (at 0.1 and 1mM) in gill, kidney, liver and brain of adult goldfish (*Carassius auratus*). Mn exposure led to different responses in fish tissues and it was dose-dependent. One mM induced an increase of lipid hydroperoxides level, superoxide dismutase (SOD) and glutathione peroxidase (GPx) activities in all tissues with the exception of SOD inhibition in the brain. Catalase (CAT) activity was inhibited in the liver and kidney, but its level was increased in the gill. Exposure to Mn 0.1mM provoked most prominent changes in the liver and did not change the enzyme activities in brain. These results strongly suggest that Mn exposure caused a generalized oxidative stress in the fish tissues and revealed an organ specific antioxidant response involving a differential modulation of the SOD, CAT and GPx activities (Vieira et al., 2011). No information of biomarker responses in fish larvae and embryos is present.

Zinc

Zinc is an essential trace metal, but many aspects of its toxicity remain unclear. Zinc usually serves as a cofactor of many enzymes (proteases, SOD, energy metabolism enzymes, etc.). It could be replaced by heavy metals, thereby making the enzymes inactive (Ercal et al., 2001). Killifish *Fundulus heteroclitus* were exposed to sublethal level (500 µg L^{-1}) of waterborne zinc for 96 h in 0% (fresh water), 10% (3.5 ppt), 30% (10.5 ppt) and 100% sea water (35 ppt). Zinc exposure clearly induced oxidative stress, and responses were qualitatively similar among different tissues. Salinity acted as a strong protective factor, with the highest levels of reactive oxygen species (ROS). High concentration of protein carbonyls and lipid peroxidation in fish tissues was observed in 0 ppt and the least values were indicated in 100% sea water (35 ppt). Increases in total oxidative scavenging capacity (TOSC) occurred at higher salinities, correlated with increases in the activities of superoxide dismutase (SOD) and glutathione-S-transferase (GST), as well as in tissue glutathione (GSH) concentrations. However, TOSC was depleted in zinc-exposed fish at 0 ppt, accompanied by decreases in SOD, GST, GSH, and also CAT activity. Thus, sublethal waterborne zinc is an oxidative stressor in fish (Loro et al., 2012a).

Selenium

Selenium is an essential element in animals. It has three levels of biological activity: (1) trace concentrations are required for normal growth and development; (2) moderate concentrations can be stored and homeostatic functions maintained; and (3) elevated concentrations can result in toxic effects. Some fish studies with selenium exposure in the water, diet, or both have reported inconsistent results: (1) reduced growth occurred in the same treatments (exposure concentration and duration) where reductions in survival occurred; (2) reduced survival occurred before reduced growth; (3) reduced growth occurred before reduced survival; or (4) no effects on growth or survival, but other pathological or reproductive effects occurred. The inconsistency between these studies was probably due to differences in species, age, exposure route and duration, selenium form and other factors. Teratogenesis is a well-documented biomarker of selenium toxicity (Hamilton, 2004; Hamilton et al., 2005).

There has been a lack of consistency of adverse and biomarker response effects from selenium exposure on either growth or survival of fish, especially in early life stages. Selenium in fish eggs is carried as part of the yolk precursor proteins, lipovitellin and phosvitin, and it is incorporated into egg immunoglobulin and vitellogenin (Hamilton et al., 2005). Selenium plays a role in antioxidant defenses as a cofactor for GPx. Organoselenium depletes the cellular antioxidant, glutathione (GSH) due to activation of organoselenides to organoselenoxides by flavin-containing monooxygenases (FMO). Since FMO tends to be induced in euryhaline fish exposed to hypersaline conditions, the developmental toxicity of salinity and organoselenium was examined in the euryhaline fish Japanese medaka (*Oryzias latipes*). The researchers measured FMO activity, GSH, and selenium concentrations in Japanese medaka embryos following a 24-h exposure to 0.05 mM l-selenomethionine (SeMet) under different saline conditions: freshwater (< 0.5 dS/m), 4.2, 6.7, and 16.8 dS/m. Concentrations of GSH and the hatch-out ratio of the SeMet-treated embryos decreased and it was dependent on salinity. While SeMet treatment led to accumulation within embryos, selenium concentrations were unaltered by salinity treatment. Compared to freshwater-exposed embryos, microsomes from embryos at 6.7 and 16.8 dS/m had enhanced oxidation of SeMet to the selenoxide (10- and 14.3-fold, respectively), which correlated with GSH depletion. The results show that increased SeMet oxidation by hypersaline conditions with subsequent GSH depletion may play an important role in the developmental toxicity of selenomethionine (Lavado et al., 2012).

Non-essential and toxic metals

Mercury

Environmental mercury and its accumulation in tissues of aquatic animals is often associated with anthropogenic pollution, but in some cases the main sources of Hg are naturally connected with geological processes in the earth and water (Turcozy et al., 2000; Rodrigues et al., 2010). Organic methylmercury and inorganic (mercurous, mercuric) forms exist in nature. Fish retained mercury in their tissues as methylmercury and its concentration increased via trophic chain from inorganic mercury in anaerobic bacteria to organic form associated with fish gills and gut. Methylmercury is the most toxic form and it is estimated for more than 95% of organic mercury in fish muscle. It may be biomagnified through all levels of the aquatic food chain as it accumulates in the top trophic level including carnivorous species (Gonul and Kucuksezgin, 2007). The increase of mercury content with age and correspondingly with body size and length is connected with slow turnover of Hg in fish. Long biological harflife causes its rapid accumulation in fish tissues (Storelli et al., 2005). In this case mercury accumulation in fish tissues plays an important role in antioxidant defense status in the organism and may modify the prooxidant-antioxidant balance of the animal. Mercury reacts with the thiol groups of GSH, which can induce GSH depletion and oxidative stress in tissues. Both organic and inorganic forms of mercury participate in the formation of ROS. Metal-induction decreases in GSH levels could be the result of direct binding of the metal to GSH through its SH group (formation of metal-SG complexes) or of enhanced oxidation of this thiol (Sevcikova et al., 2011). Mercury inhibits of mitochondrial oxidative phosphoriyation, damage calcium homeostasis, induces lipid peroxidation and alters antioxidant enzyme activities (Ercal et al., 2001).

Acute toxicity tests of mercuric chloride to Japanese flounder (*Paralichthys olivaceus*) indicated that the 48-h LC_{50} values of mercury to the embryos and larvae were 48.1 (32.8–63.6) and 99.4 (72.9–147.0) μgL^{-1}, respectively. Mercury could cause low hatching success, delayed hatching process, reduced growth at concentrations $\geq 20 \mu gL^{-1}$, and reduced survival and higher morphological malformations at concentrations $\geq 40 \mu gL^{-1}$ in the embryos and larvae (Huang et al., 2010a). The authors documented that the hatching, survival, growth and antioxidants of the flounder were sensitive to the highest mercury concentrations and could thereby serve as potential biomarker for evaluating mercury contamination in the aquatic environment. Increasing mercury concentration led to increased mercury bioaccumulation and reduced flounder growth. The authors suggest that flounder larvae and juveniles have the potential to manipulate the levels of antioxidants such as SOD, CAT and GSH activities, which protect flounder from oxidative stress induced by mercury exposure. These antioxidants could serve as biomarkers of mercury contamination in the aquatic environment and biota (Huang et al., 2010b).

Our results have shown that the response of antioxidant enzymes in the developing embryos of the round goby *Neogobius melanostomus* on different concentrations of mercury chloride was not uniform and depended on metal concentration (Rudneva, 1993). The increase of mercury concentration in the water resulted in the decrease of protein concentration in fish embryo more than 80% compared to the control (r = 0.99). The fluctuations of protein concentration and enzyme activities in developing embryos (stage VI) are present in Fig. 4.2.

We suggest, that the most sensitive enzyme to mercury impact was CAT because the increase of Hg concentration resulted in the increase of CAT activity more than three orders. This antioxidant enzyme could serve as biomarker of mercury contamination in the aquatic environment. The response of fish embryos and larvae biomarkers on mercury exposure are summarized in Table 4.1.

Study of the impact of mercury exposure on goldfish (*Carassius auratus*) embryos noted the dynamic characteristics of chemical parameters. Day-old embryos were exposed to different Hg^{2+} concentrations (0, 0.2, 1, 5, and 10 µg/L). The responses of acid phosphatase (ACP) and alkaline phosphatase (AKP) to mercury exposure were presented in dose-dependent and time-dependent manners. The enzyme activities were significantly induced with

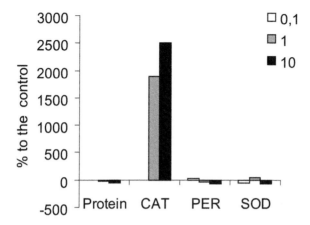

Fig. 4.2. Protein concentration and antioxidant enzyme activities in round goby *N. melanostomus* embryos (stage VI) effected by mercury chloride. 0.1, 1.0 and 10.0–mercury concentration, µg HgCl L⁻¹ respectively, time of exposure was 1 h.

increased concentrations and extended exposure (at 5 µg/L after 72 h and 10 µg/L after 48 h; $p < 0.05$ or $p < 0.01$). Lysozyme (LSZ) was not sensitive to lower Hg²⁺ concentrations, whereas LSZ significantly increased at higher concentrations and longer exposure (at 5 µg/L at 120 h and 10 µg/L after 72 h; $p < 0.05$ or $p < 0.01$). CAT activities were significantly inhibited at different periods of embryonic development, particularly at 5 and 10 µg/L ($p < 0.05$ or $p < 0.01$). Reduced CAT activities were observed at 72, 96, and 120 h at 1 µg/L ($p < 0.05$ or $p < 0.01$), whereas a decline at 0.2 µg/L was evident at 96 h ($p < 0.01$). MDA content significantly increased at various stages of embryonic development, particularly at 10 µg/L ($p < 0.05$ or $p < 0.01$), and increased further at 72, 96, and 120 h at 5 µg/L ($p < 0.05$ or $p < 0.01$). At 96 h, MDA content was only increased by exposure to 0.2 and 1 µg/L ($p < 0.01$). The activities of ACP, AKP, and LSZ increased at 120 h in contrast to 96 h ($p < 0.05$ or $p < 0.01$). Therefore, 96 h is an important shifting period of embryonic development because the activity of enzyme has been enhanced at this time. Thus, the increased ACP, AKP, and LSZ activities revealed an enhanced ability of the embryo to synthesize more enzymes and attenuate mercury damage. CAT activity negatively correlated with MDA accumulation. The enhanced enzyme activities after specific embryonic stages are used to strengthen the ability to cope with mercury stress and attenuate mercury damage. The biochemical parameters, except LSZ, exhibited sensitivity to mercury, suggesting that they may act as potential biomarkers in assessing the environmental mercury risk on *C. auratus* embryos (Kong et al., 2012).

Table 4.1. Response of fish biomarkers on mercury impact.

Fish species and life stages	Concentration and exposure	Response	References
Japanese flounder (*Paralichthys olivaceus*) embryos and larvae	Mercuric chloride, 10 days of 0–10µgHg^{2+}L^{-1}	SOD and CAT activities, reduced glutathione (GSH) and MDA contents of the larvae were significantly increased, while (GST) was decreased	Huang et al., 2010a
Japanese flounder (*Paralichthys olivaceus*)	0–10 µg Hg^{2+}L^{-1} for 80 days	SOD and CAT activities at the three developmental stages were increased with increasing mercury concentration, while GST activity did not significantly vary	Huang et al., 2010b
metamorphosing larvae (18 days post hatching, dph settling larvae (33 dph) juveniles (78 dph)		Glutathione (GSH) content was elevated, MDA content did not change MDA content significantly increased Glutathione (GSH) content decreased	
Round goby, *Neogobius melanostomus*	0.1 µg HgCl L^{-1}	Protein content, CAT and SOD activities decreased, PER activity increased	Rudneva, 1993
VI stage embryo	1 µg HgCl L^{-1}	Protein content, PER activity decreased, CAT and SOD activity increased	
	10 µg HgCl L^{-1}	Protein content, PER and SOD activity decreased, CAT activity increased	
Carassius auratus day –old embryos	Hg^{2+}, 0.2, 1, 5, and 10 µg/L for 24 h		Kong et al., 2012

Lead

Lead is a major environmental pollutant. Lead can induce oxidative damage through direct effects on the cell membrane, interaction between lead and hemoglobin, which increases the auto-oxidation of hemoglobin, auto-oxidized δ-aminolevulinic acid, interaction with GR, or through the formation of complexes with selenium, which decreases GPx activity (Sevcikova et al., 2011). Lipid peroxidation in the brain of lead-exposed fish was observed and several mechanisms are proposed for lead-induced oxidative stress. They are the following: direct effects of lead on cell membranes via deterioration of their components; lead-hemoglobin interactions, caused hemolisys and lipid peroxidation; δ-aminolevulinic acid (δ-ALA)-induced generation of reactive oxygen species and effect of lead on the antioxidant defense systems of cells via blocking SH-groups of antioxidant enzymes (Ercal et al., 2001). The information of biomarker response in fish early developmental stages on lead exposure is very scarce.

Cadmium

Cadmium is a toxic element with unknown function. It is nondegradable and cumulative toxicant. The mechanisms responsible for Cd-induced toxicity may be multifactorial and they involve the following: adverse effects of Cd on cellular defense systems and thiol groups content; enhancement of lipid peroxidation and deleterious effects on cellular enzymes (Ercal et al., 2001). Cd can damage gills and decrease the activity of gill Ca^{2+}-ATP-ase, which leads to fish hypocalcemia and can result in skeletal deformities and disturbed Ca balance. Cadmium has adverse effects on growth, reproduction, respiratory functions and osmoregulation (Wu et al., 2007). Cadmium has been well characterized in terms of its absorption, tissue distribution, mechanism of action, and elimination in many adult vertebrate species, whereas the potential embryonic effects of this heavy metal are poorly understood. Although Cd exposure can give rise to a variety of developmental defects, it has proven difficult to identify toxin-sensitive cells among the wide range of cell populations in a developing embryo (Blechinger et al., 2002).

Cadmium does not generate ROS directly, but can alter GSH levels and influence cell thiol status, inducing the expression of metallothioneins in the liver. Changes in GSH and MTs can lead to LPO of the cell membrane. Cadmium enters the electron transport chain in mitochondria, leading to accumulation of unstable semiubiquinones which donate electrons and create superoxide radicals. Cadmium also affects antioxidant enzymes, especially SOD and CAT, and is able to displace copper and iron in various proteins, freeing these metals to then participate in the Fenton reaction. Metallothioneins play a major role in the detoxification of cadmium, and this process is clearly organ-specific. The effects of cadmium exposure on GSH levels and antioxidant enzyme activities vary with fish species, duration of exposure, and the chemical involved (Sevcikova et al., 2011; Wu et al., 2007). MT expression in newly hatched tilapia larvae can be stimulated with dose- and time-dependent patterns after 48 hours' exposure to 35µg/L Cd^{2+}. In contrast, 3-day-old larvae receiving the same treatment revealed serious impairments in physiologic performance, an approximately 10-fold increase in Cd^{2+} accumulation, and a 10% decrease in Ca^{2+} content, but no significant change in MT content was found compared to the control (Wu et al., 2000). The next study of these authors suggested that MT content was not significantly higher than the controls at 48 hours exposure to 40 µg/L Cd^{2+}, but it increased by 4.5-fold on treatment with 40 µg/L Cd^{2+} + 2mM Ca^{2+} compared to the controls. Therefore, in tilapia H3 larvae 40 µg/L Cd^2 may be a high dose that may cause detrimental effects and thus inhibit MT expression. The addition of Ca^{2+} to the Cd^{2+} medium caused stimulation of MT expression in larvae. Thus, the additional Ca^{2+} may decrease Cd^{2+} toxicity to larvae and allow the organisms to express more MT. However, dose-response of MT expression appears only when metals do not introduce detrimental effects to the physiologic function of larvae (Wu et al., 2007).

Cd exposure affected oxidative biomarker in early life stages of Japanese flounder, *Paralichthys olivaceus*. Fish were exposed to waterborne Cd (0–48 microg L^{-1}) from embryonic to juvenile stages for 80 days. Growth, Cd accumulation, activities of SOD, CAT, GST, and levels of glutathione (GSH) and lipid peroxidation (LPO) were investigated at three developmental stages. Flounder growth decreased and Cd accumulation increased with increasing Cd concentration. In metamorphosing larvae, CAT and SOD activities were inhibited and GSH level was elevated, while LPO was enhanced by increasing Cd concentrations. CAT and GST activities of settling larvae were inhibited but GSH level was elevated at high Cd concentrations. In juveniles, SOD activity and LPO levels were increased but GST activity decreased as Cd concentration increased. Antioxidants in flounder at early life stages were able to develop ductile responses to defend against oxidative stress, but LPO fatally occurred due to Cd exposure. These biochemical parameters could be used as effective biomarkers for evaluating Cd contamination and toxicity in marine environments: CAT, SOD, GSH, and LPO for metamorphosing stage; CAT, GSH, and GST for settling stage; and SOD, GST, and LPO for juvenile stage (Cao et al., 2010). The authors also demonstrated the tissue-specific accumulation of cadmium and its effects on antioxidative responses in Japanese flounder juveniles. Following Cd exposure for 28 d, accumulation of Cd in fish was dose-dependent and tissue-specific, with the greatest accumulation in the liver, followed by the kidney, gill, and muscle. Although the gill and liver mounted active antioxidant responses at ≥ 4 mg L^{-1} Cd including a decrease in glutathione level and GST and GPx activities, the antioxidant response failed to prevent lipid peroxidation induction in these organs. In the kidney, increased GPx and GST activities and decreased SOD activity were observed in fish exposed to high Cd concentrations, but LPO levels did not significantly differ among the exposure concentrations. The gill was most sensitive to Cd exposure. The gill was the most sensitive to oxidative damage, followed by the liver; the kidney was the least affected tissue (Cao et al., 2011).

Cd as well as a variety of cytotoxic agents have been shown to up-regulate hsp70, and hsp70 expression has been examined as a potential marker in toxicologic screening for a number of vertebrate species, including fish. The establishment of an *in vivo* system that uses hsp70 gene activation as a measure of cadmium toxicity in living early larvae of transgenic zebrafish carrying a stably integrated hsp70-enhanced green fluorescent protein (eGFP) reporter gene. eGFP expression in this strain of fish acts as an accurate and reproducible indicator of cell-specific induction of hsp70 gene expression. The transgene fish responds in a dose-dependent manner at concentrations similar to those observed for morphologic indicators of early-life stage toxicity and was sensitive enough to detect cadmium at doses below the median combined adverse effect concentration and the median lethal concentration. The stable nature of this transgenic line should allow for extremely rapid and reproducible toxicologic profiling of embryos and larvae throughout development. The patterns of expression for hsp70 and hsp70-eGFP were investigated in larvae following acute pulse (3hr) exposures to cadmium. The patterns of expression for hsp70 and hsp70-eGFP were dose-dependent and tissue specific. Cadmium induced expression of hsp70 and hsp70-eGFP in the skin, gills, olfactory rosette, liver, and kidney, all of which are either accumulators or target organs of cadmium in fish (Blechinger et al., 2002).

Arsenicum

Feeding habits have important influence on arsenic levels in marine fish: animal feeding of algae and crustacea appear to retain higher arsenic concentrations than piscivorous forms. Most marine animals have only limited ability to accumulate arsenic from seawater and its concentrations in the fish depend on their food and trophic position in the local food chain. The element assimilated from the food is presented in the tissues in organic form, whereas that accumulated from water remains inorganic and it is excreted rapidly or sequestered in non-living tissue compartments (Neff, 1997). Arsenic in marine organisms is found in the fat or water-soluble forms such as arsenobetaine which is less toxic than inorganic form (Turoczy et al., 2000). It usually represents 50% to more than 95% of the total arsenic in tissues of marine invertebrates and fish (Neff, 1997). Arsenic also induced expression of *hsp70* in the liver, gills, olfactory rosette and skin in the larvae of zebrafish however cadmium was much more toxic than arsenic (Blechinger et al., 2002).

Arsenite (III) and arsenate (V) are inorganic forms that can be methylated. The trivalent arsenite is biologically more active than pentavalent arsenate. Glutathione plays a key role in the cell redox status induced by arsenic. Glutathione is an electron donor in the reduction of arsenate to arsenite. Arsenic cell metabolism generates ROS, although the mechanisms are not clear. Reactive nitrogen species are also involved in oxidative damage associated with arsenic (Bhattacharya and Bhattacharya, 2007). Arsenic-induced free radical formation and damage on cellular antioxidant defense systems were also documented (Ercal et al., 2001).

Nickel

Toxic and carcinogenic effects of nickel compounds are suggested to result from nickel-mediated oxidative damage to macromolecules and/ or inhibition of cellular antioxidant defenses (Pane et al., 2003). The waterborne Ni^{2+} (10, 25 and 50 mg/L) affected blood and blood-producing tissues (kidney and spleen) of adult goldfish and resulted in damage of immune system and lymphopenia. Ni accumulation increased renal iron content (by 49–78%) and resulted in elevated lipid peroxides (by 29%) and protein carbonyl content (by 274–278%), accompanied by suppression of the activities of SOD (by 50–53%), GPx (15–45%), GR (31–37%) and glucose-6-phosphate dehydrogenase (20–44%), indicating development of oxidative stress in kidney. In contrast to kidney, in spleen the activation of GPx (by 34–118%), GST (by 41–216%) and GR (by 47%), as well as constant levels of low molecular mass thiols and metals together with enhanced activity of glucose-6-phosphate dehydrogenase (by 41–94%) speaks for a powerful antioxidant potential that counteracts Ni-induced ROS production. As Ni accumulation in this organ was negligible, Ni-toxicity in spleen may be minimized by efficient exclusion of this otherwise toxic metal (Kubrak et al., 2012). There is a lack of information of Ni toxic effects and biomarker response in fish early life.

Silver

AgNPs are widely employed in commercial products, and are thus inevitably released into the aquatic environment. AgNPs could induce toxicological effects on fish embryonic development. AgNP treatment at the concentration of 62.5–1000 µg/L caused significant increase in retarded development and abnormalities in early-life stages of medaka fish (*Oryzias latipes*). Destruction of the surface ornamentation and egg envelope was observed at a higher AgNP concentration (≥ 125 µg/L). A dose-dependent increase in lactate dehydrogenase (LDH) activity, an indicator of anaerobic metabolism, and SOD activity was noted in the treated embryos. In contrast, the total reduced glutathione level decreased. A high thiobarbituric acid reactive substance concentration in embryos was generated upon AgNP exposure from day 1 to day 7 postfertilisation. The biochemical parameters suggested that oxidative stress was induced by the AgNPs. A dose-dependent reduction in ROS and $^1O_2$ generation upon high AgNP exposure (≥ 250 µg/L) was shown. Although the morphological damages induced by the AgNPs were irreversible, restorable antioxidant defenses were noted in the well-developed embryos. The authors supported the idea that the stage of morphogenesis and organogenesis is a critical window to chemical exposure or environmental stress. Overall, the results suggested that hypoxia, disturbed egg chorion, and oxidative stress are closely associated with AgNP toxicity in embryonic fish (Wu and Zhou, 2012).

Platinum

The platinum group metals (PGMs) platinum (Pt), palladium (Pd), and rhodium (Rh) are used in automobile catalytic converters, from which they have been emitted into the environment to an increasing degree during the last 20 years. Studies determining the effects of PGMs on organisms are extremely rare. Effects of various concentrations of $PtCl_2$ (0.1, 1, 10, 50 and 100 µg/L) were investigated with respect to the induction of hsp70 and histopathological alterations in the zebrafish, *Danio rerio*. Histopathological investigations revealed effects of Pt on fish, which varied between slight and strong cellular reactions, depending on the $PtCl_2$ concentration. The hsp70 level was significantly elevated at 100 µg/L $PtCl_2$ in *D. rerio* (Osterauer et al., 2010).

Hence, the above mentioned studies document that metals induced toxic effects in fish early life which associated with the oxidative stress. The response of fish early developing stages to oxidative damage after acute and chronic metal exposure is evident under laboratory conditions. However, the biomarkers are diversely influenced by metals. The data of fish response on metal toxicity are contradicted: increases and decreases in enzyme activities and enhanced and reduced levels of low molecular weight scavengers have been described after metal exposure. A specific biomarker of oxidative stress caused by metals does not exist, and for that reason a complex approach should be taken. Metallothioneins seem to be a suitable biomarker of metal exposure under laboratory conditions. The sensitivity of fish embryos to metals is stage-dependent and it depends on the function of the organism's ability to metabolize the chemical or activate cellular stress responses and repair mechanisms.

4.1.2 Organic pollutants effects

Crude oil and its fractions

The toxicity of oil fractions and products and response of fish early life stages are also varied depending on experimental duration. There are several studies of oil effects on adult fish using the battery of biomarkers while the information of the response of early life stages of fish and invertebrates is scarce. Generally the investigators describe the hatching process, mortality and abnormalities of development of the embryos and larvae under oil impact (Beyer et al., 2001; Pauka et al., 2011). Zebrafish embryos were investigated to assess oil toxicity. For instance, they were used to evaluate the potential toxicological differences between unrefined crude and residual fuel oils, and test the effects of sunlight as an additional stressor. Modern bunker fuels contain residual oils that are the highly processed and chemically distinct remains of the crude oil refinement process. Using mechanically dispersed oil preparations, the embryotoxicity of two bunker oils was compared to a standard crude oil from the Alaska North Slope. In the absence of sunlight, all three oils produced the stereotypical cardiac toxicity that has been linked to the fraction of tricyclic aromatic compounds in an oil mixture. However, the cardiotoxicity of bunker oils did not correlate strictly with the concentrations of tricyclic compounds. When embryos were sequentially exposed to oil and natural sunlight, the bunker oils produced a rapid onset cell-lethal toxicity not observed with crude oil. The much higher phototoxic potential of chemically distinct bunker oils observed here suggests that this mode of action should be considered in the assessment of bunker oil spill impacts, and indicates the need for a broader approach to understand the aquatic toxicity of different oils (Hatlen et al., 2010).

The toxic effects of water-soluble fraction (WSF) of crude oil (AP127, Petobras Campos Basin, Brasil) were studied during early development of zebrafish during 96 h and its biotransformation in juvenile fish. Reduced heartbeat rate, weak pigmentation, tail defects, and embryo mortality were observed for all the tested concentrations of the WSF (15%, 33% and 50%). Activities of the biotransformation enzymes were induced at the highest concentrations, showing that these enzymes played a role in its elimination. In juvenile fish exposed to 33% and 50% concentration of WSF of crude oil the activities of EROD and GST increased significantly compared to the control. The authors documented that the crude oil WSF altered the normal embryonic development of fish and induced biomarkers of its biotransformation (Pauka et al., 2011).

The authors studied the toxicity of various oil fractions on early life stages of some Black Sea fish species and their biomarker response. Toxic effects of oil (direct exposure, group I) and its washout (indirect exposure, group II) which were collected from marine sediments in Sevastopol Bay

(Black Sea, Ukraine) were evaluated in embryos of *N. melanostomus* (stages V and VI) and hatching larvae (Rudneva, 1998; Chesalina and Rudneva, 1998; Rudneva and Shaida, 2011). Biomarker (lipid peoxidation and antioxidant enzyme activities) responses were tested (Fig 4.3; 4.4). Lipid peroxidation

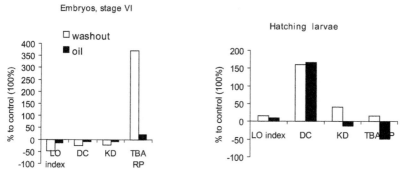

Fig. 4.3. The effect of heavy oil fractions and their washout on lipid peroxidation of *N. melanostomus* (embryos stage VI) and hatching larvae. LO index—lipid oxidation index, DC—dien conjugates, KD—ketodiens, TBA RP—Thiobarbituric reactive products (time of exposure 96h, T = 12.5–13.1°C).

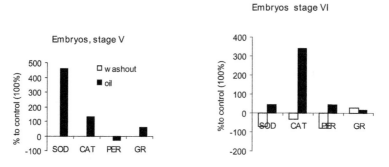

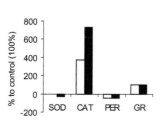

Fig. 4.4. The effect of heavy oil fractions and their washout on antioxidant enzyme activities of *N. melanostomus* (embryos stages V and VI) and hatching larvae. SOD—superoxide dismutase, CAT—catalase, PER—peroxidase, GR—glutathione reductase (time of exposure 96h, T = 12.5–13.1°C).

parameters varied insignificantly in both experimental groups of embryos as compared to the control with the exception of TBA-reactive products whose level increased significantly (p < 0.01). In hatching larvae the similar trend was observed in dien conjugates level which was higher in treated fish than those in control group.

The response of antioxidant enzyme activities to oil exposure depended on fish developmental stage and durations of experiment (Fig. 4.4). SOD, CAT and GR activities were higher in embryos stages V and VI exposed directly in heavy oil fractions (group I). Tested parameters varied insignificantly in embryos in group II (indirect exposure) and tended to decrease with the exception of GR. In hatching larvae in both experimental groups CAT activity was greater than that in control to 340–730%, the similar trend was observed in GR activity (90–100% to control) while SOD and PER levels varied insignificantly.

Therefore, it could be concluded that the direct exposure of heavy oil fractions and their "washout" (indirect effect) were not uniform in fish embryos and hatching larvae in both tested groups. However, in general, direct and indirect effects of oil caused the increase of the final products of LPO (TBA-reactive products) in embryos and intermediate compounds concentration in larvae. Thus, in larvae "late response" was observed as compared to embryos which can be explained by the increase of antioxidant enzyme activities in larvae than in embryos (see Chapter 3).

Selected biomarker effects were analysed in *N. melanostomus* embryos stage V and VI exposed in various concentrations of of waterborne mixtures of crude oil (mazut and diesel fuel) (Fig. 4.5). SOD and CAT activities did not respond in embryo stage VI exposed in diesel fuel 1 ml L^{-1} while PER and GR levels increased significantly. Antioxidant enzymes activity in embryos V stage decreased in low and medium concentrations of mazut with the exception of CAT and GR whose levels increased at concentration 0.1 ml L^{-1}. In embryos stage VI the trends of PER and GR were similar and characterized a significant increase at low and medium concentrations of mazut and decrease at high concentration. SOD and CAT activities varied insignificantly as compared to PER and GR. Therefore, the fluctuations of antioxidant enzyme activities depended on kind of toxicant, its concentrations and developmental stage of fish.

The response of embryos stage VI of *Lypophys pavo* exposed to various concentrations of water soluble oil fractions is present in Fig. 4.6.

The toxicity of diesel fuel was greater for *L. pavo* embryos as compared to mazut and the lethal concentration of diesel fuel was 0.1 ml L^{-1}. SOD and CAT activities increased progressively at all tested concentrations of diesel fuel while PER activity varied insignificantly and GR level dropped at 0.01 and 0.1 ml L^{-1} and increased at 0.05 ml L^{-1}. On the contrary, SOD and GR activities decreased at low concentration of mazut and increased with the

Mazut, embryo stage V

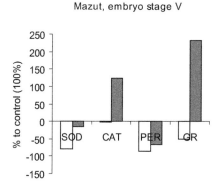

Mazut, embryo, stage VI

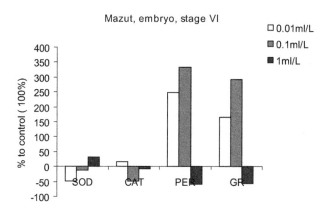

Diesel fuel, embryo, stage VI

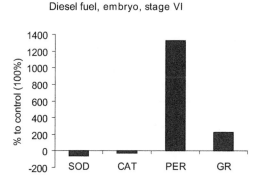

Fig. 4.5. The effect of water soluble oil fractions mazut and diesel fuel on antioxidant enzyme activities of *N. melanostomus* (embryos stages V and VI), T = 12.5–13.1°C.

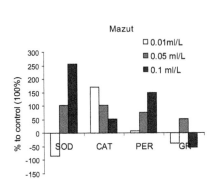

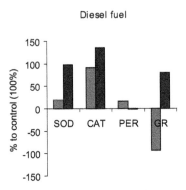

Fig. 4.6. Effects of diesel fuel and mazut on antioxidant enzyme activities of *Lypophys pavo* embryo stage VI, 0.01, 0.05 and 0.1 ml/L—concentration of diesel fuel and mazut respectively, time of exposure was 1 day, T = 12.5–13.1°C.

increase in toxicant level, PER and CAT elevated in all tested concentrations. Thus the response of antioxidant enzyme activities was not uniform and depended upon oil concentrations and kind of toxicant.

We also studied the effect of various concentrations of waterborne mixtures of crude oil in early life stages of Black Sea fish species. Our studies have shown that diesel fuel toxicity was greater than *Atherina* larvae because at concentration of 5 ml L⁻¹ fish died. Antioxidant enzyme activities were decreased in larvae incubated in diesel fuel at concentration of 2.5 ml L⁻¹ with the exception of PER level which was greater than in control group (Fig. 4.7). We could propose that mazut caused the toxic response in larvae antioxidant enzymes and inhibited enzymatic activity. It agrees with the results of our next studies of toxic effects of mazut on *Atherina* larvae antioxidant enzyme status (Fig. 4.8).

We observed that in high concentration of mazut (5 ml L⁻¹) antioxidant enzyme levels decreased significantly while in low concentration (2.5 ml L⁻¹) the enzymatic activity elevated progressively during the time of exposure (with the exception of SOD). We could propose that PER plays an important

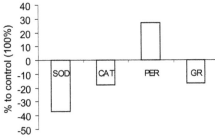

Fig. 4.7. Effects of diesel fuel (2.5 ml L⁻¹) on antioxidant enzyme activities of *Atherina* larvae, time of exposure was 1 day.

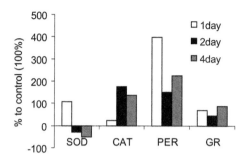

Concentration 2.5 ml per L

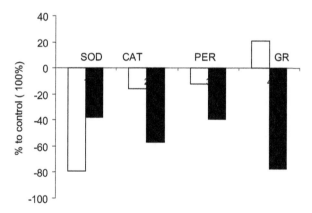

Concentration 5 ml per L

Fig. 4.8. Effects of mazut (2.5 ml L⁻¹ and 5 ml L⁻¹) on antioxidant enzyme activities of *Atherina* larvae, time of exposure 1, 2 and 4 days.

role in detoxification process of WSF because its activity in treated larvae was higher than other enzymes as compared to control.

We investigated the effect of diesel fuel (concentration 2.5 and 5 ml L⁻¹) on *Liza saliens* fry and we found that the concentration of 5 ml L⁻¹ was lethal to fish because they died during 20 h (Chesalina et al., 2000). The response of antioxidant enzyme activities in fish exposed at 2.5 ml L⁻¹ of diesel fuel during 4 days was not uniform. On 1st day of exposure the enzymes level was dropped as compared to control, then it increased (with the exception of PER) in day 2. On day 4 SOD and PER activities again decreased, CAT activity elevated and GR was similar as control value (Fig. 4.9). Therefore, the decrease of antioxidant enzyme levels during four days of exposure could explain the toxicity of diesel fuel which suppressed their activities and led to the mortality of *L. saliens* fry.

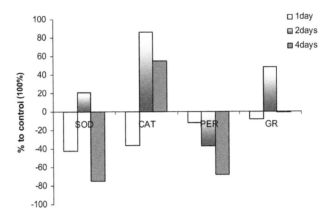

Fig. 4.9. Effects of diesel fuel (2.5 ml L^{-1}) on antioxidant enzyme activities of *Liza saliens* fry-time of exposure 1, 2 and 4 days.

We also found that the toxicity of mazut was lower as compared to diesel fuel and the lethal effect of fish fry exposed in mazut (concentration 5 ml L^{-1}) as observed during the first day of exposure (Fig. 4.10).

Activities of SOD and CAT tended to decrease as compared to control during four days of exposure, then CAT level increased at the concentration of 2.5 ml L^{-1}. PER and GR activities progressively elevated at the first day of exposure then dropped and elevated at day 5. Enzymatic activity decreased at concentration of 5 ml L^{-1} (with the exception of CAT) and the mortality of fish fry was noted (Rudneva et al., 2000). We could conclude that the response of antioxidant enzyme system includes several phases: at the beginning of intoxication the induction of antioxidant enzyme activities play an important role in defense of the organism against toxic agents, then the toxicants accumulate in tissues and suppress the enzymatic activity. The third phase of toxicity is characterized by the induction of defense mechanisms and resistance of the organism to intoxication. On the other hand, at high concentration of toxicant (5 ml L^{-1}) antioxidant enzyme activities were inhibited at 1st day of fish exposure and caused the mortality of fish embryos.

The obtained results showed that all examined stressors led negative biological effects on fish embryos, larvae and fry. They resulted in increased mortality and decreased survival. Toxicants induced lipid peroxidation processes in larvae and caused the increase of free radical production. The responses of antioxidant enzyme system varied and depended upon:

- kinds of oil components, their concentrations and time of exposure;
- fish developmental stage: larvae were more sensitive than embryos and the response of embryos depended on eggshell thickness (Table 4.2): the great thickness of shell protects the developing embryo against

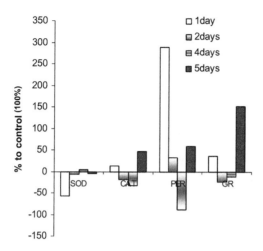

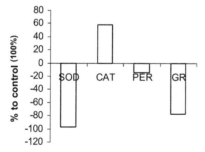

Fig. 4.10. Effects of mazut (2.5 ml L^{-1} and 5 ml L^{-1}) on antioxidant enzyme activities of *Liza saliens* fry-time of exposure 1, 2, 4 and 5 days.

toxic impact of oil and the response of antioxidant enzyme activities was less sensitive as compared to the embryos with thin eggshell;

• fish species individual peculiarities: *Atherina hepsetus* larvae were more resistant than *Liza aurata* larvae.

Thus all studied parameters reflect negative effects of environmental stress caused by oil pollution on fish embryos, larvae and fry and can be used as biomarkers in biomonitoring of coastal areas and for assessment of their health. However, the response of tested biomarkers was not uniform and depended upon both fish developmental stages, kinds of oil products

Table 4.2. Response of antioxidant enzymes in fish embryos exposed to various oil concentrations.

Fish species	Eggshell thickness, µm	The lowest oil concentrations which induces the changes of antioxidant enzyme activities, ml L^{-1}	Range (min-max) of enzymatic activities response, %
Psetta maxima maeotica	3	10^{-6}	$-40 \div +1300$
Gobiidae (N. melanostomus)	4	10^{-5}	$-40 \div +600$
Blennidae (L. pavo)	5	10^{-5}	$-60 \div +160$

and duration of experiment. Further investigations and the additional biomarkers are needed for the evaluation of the general trends of oil toxicity in fish early developmental stages.

PCB, TBBP, PAH and other organic compounds

The polychlorinated biphenils (PCB) and Cl-containing persistent organic compounds are highly distributed environmental pollutants due to their extensive application in different kinds of human activity. As other polyaromatic hydrocarbons (PAH) they sequester in sediments in aquatic ecosystems and pose a hazard to the aquatic organisms especially benthic forms that are exposed to them directly or via food chains (Maenpaa et al., 2004). These xenobiotics impact marine organisms and cause damage of cellular and membrane components, biological molecules and result in malformation of the developing eggs and embryos of fish and invertebrates.

We observed the effects of PCB (Arochlor 1254) in the concentrations of 0.0001 and 0.001 µg·1^{-1} on the heat production of developing eggs IV and VI stages and hatching larvae of round goby *N. melanostomus* Pallas (Rudneva and Shaida, 2012). No significant differences were observed in metabolic rate in fish developing embryos in all examined stages, exposed in PCB in both concentrations during 50 h at the temperature +20°C (Fig. 4.11). However, heat production of the hatching larvae exposed to the marine water containing PCB in both tested concentrations was significantly lower ($p < 0.01$) as compared to intact fish.

The similar trend was indicated in *A. hepsetus* larvae, incubated in solvent (hexane 1 ml.L^{-1}) and in the mixture hexane + PCB (0.001µg·1^{-1}) (Fig. 4.12). Heat production of the animals treated by both chemicals was significantly lower than the metabolic rate of the control fish.

Thus, our finding demonstrated no significant differences between intact and treated fish eggs heat production in tested concentrations of the PCB. However, larvae were more sensitive to toxicants and they showed significant decrease ($p < 0.01$) of the metabolic rate in both tested PCB concentrations and solvent. In the case of *N. melanostomus* larvae heat

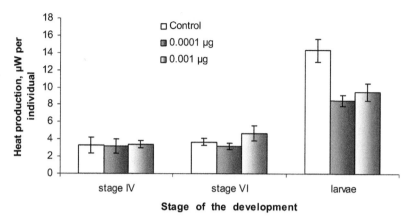

Fig. 4.11. Metabolic rate of *N. melanostomus* developing embryos (stage IV and VI) and hatching larvae exposed to PCB in the concentrations of 0.0001 and 0.001 μg·l⁻¹ (n = 3, Mean+ SD). Time of monitoring in the Monitor of biological activity was 50h at the temperature +20°C.

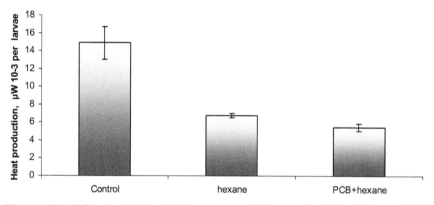

Fig. 4.12. Metabolic rate of *A. hepsetus* larvae exposed to hexane (1 ml.l⁻¹) and hexane +PCB (0.001μg·l⁻¹) (n = 3, Mean+ SD). Time of the monitoring in the Monitor of biological activity was 20h at the temperature +20°C.

production was dropped to 35–40% as compared to the values of control animals while in treated *A. hepsetus* it was 54% in the case of solvent exposure and 63% in the case of the mixture hexane+PCB.

The obtained results show that larvae are more sensitive to PCB toxicity than eggs. *A. hepsetus* larvae heat production decreased more than *N. melanostomus* larvae. The general trend of metabolic rate decrease could be explained by the imbalance in the processes of energy generation and utilization and modification of the metabolic pathways. Toxicity reduces metabolic rate which was demonstrated in moving activity, decrease of enzyme activities, blood vessel pulse in some invertebrates (Maenpaa et al., 2004). The ability to reduce metabolic rate during the exposure to environmental stress, termed metabolic rate suppression is an important component to enhanced survival in many organisms. Metabolic rate suppression can be achieved through modifications to behavior, physiology, and cellular biochemistry, all of which act to reduce whole organisms energy expenditure (Richards, 2010).

The results obtained on PCP-treated salmon eggs (0.992 μmol$\cdot$l^{-1}) demonstrated significant increase in the heat dissipation when compared with the ethanol (solvent) control and with intact group. In the higher concentrations the heat production decreased, which may indicate that PCP exposure was high enough to kill the fish embryo. PCP is a toxicant for uncoupling of phosphorilation and low concentrations of the PCP uncouples oxidative phosphorylation thus disrupting the regulation of energy metabolism by inhibiting the formation of ATP (Maenpaa et al., 2004). Penttinen and Kukkonen (2006) showed significant increase of heat production of the alevins of *Salmo salar* exposed to pentachlorphenol (PCP) in different concentrations. The output-mediated link between tissue residues concentration and heat output was demonstrated and the high correlation between bioaccumulation of the PCP and exotoxicity was revealed. Larvae were more sensitive to the toxicants and embryo accumulating high amount of chemical in the late developmental stage possibly intensified the bioenergetic effect of PCP.

In the case of PCB-treated fish eggs and larvae we observed that larvae are more sensitive to the toxicant impact than the eggs. The similar effects were documented by the other authors as well (Maenpaa et al., 2004; Penttinen and Kukkonen, 2006). In the case of fish egg the eggshell is the main barrier for xenobiotics uptake into egg. The larvae are more sensitive to chemicals impact, because they have no shell. The uptake rate of chemicals was higher in the developmental stage approach in hatching which probably suggests increasing interaction of egg with surrounding environment and that eggshell supposed to forfeit its role as a protective barrier along the development of the egg (Maenpaa et al., 2004). The strengthening of toxic effects in larvae may also be related to the increasing intake of exogenous substrates from the environment.

Our results documented that concentration of PCB 1500 ng L^{-1} is toxic for *Atherina* larvae and fish died during 2 h from the beginning of exposure. (Rudneva et al., 2003). The changes of antioxidant enzyme activities in fish larvae exposed in different concentration of PCB is present in Table. 4.3.

No changes in enzyme antioxidant activities were shown in larvae exposed to 0.1–1.0 ng L^{-1} PCB. SOD activity increased to 51% at the concentration of 1500 ng L^{-1}. During 24 h CAT activity enhanced at the concentration of 50 ng L^{-1}. Increase of PCB concentration to 500 ng L^{-1} enzyme activity dropped to the values of the control and then enhanced at the concentration of 1500 ng L^{-1}. Significant increase of GR activity was

Table. 4.3. Antioxidant enzyme activities in Atherina larvae exposed to different concentrations of PCB (mg protein^{-1} min^{-1}, mean ± SEM, n = 3).

PCB concentration, ng L^{-1}	SOD, arbitrary units	CAT, mgH$_2$O$_2$ X 10^{-1}	GR, nmol NADPH
Control	43.35 ± 10.67	1.37 ± 0.191.	0.77 ± 0.21
0.1	47.74 ± 11.32	1.13 ± 0.36	1.20 ± 0.27
1.0	42.18 ± 10.28	1.57 ± 0.20	1.20 ± 0.30
50.0	42.82 ± 17.41	3.03 ± 0.41*	3.21 ± 0.71*
500.0	40.24 ± 15.39	1.75 ± 0.12	2.73 ± 1.20*
1500.0	65.11 ± 17.40	3.47 ± 0.79*	2.93 ± 0.81*

*Significant differences between control and experimental groups (p < 0.01)

indicated in larvae at the concentration of 50 ng L^{-1} and higher. Thus, CAT and GR activity were the most sensitive to PCB exposure and their response was shown at the concentration of 50 ng L^{-1}. In addition, we noted the variations of protein electrophoretic composition of fish larvae treated by PCB at the concentration of 1 ng L^{-1}. The decrease of electrophoretic components number and changes in their mobility was documented. We proposed that PCB may modify protein properties and their physical and chemical characteristics. Similar changes in adult fish exposed to different PCB concentrations have been described earlier. The decrease of SH-groups content in fish serum upto 40–80% and the variations of protein components number were noted (Rudneva and Jerko, 1993). Thus, protein composition and its modifications as the response on toxicants exposure may be used as a good biomarker.

Therefore, obtained results demonstrated that the concentration of 1500 ng l^{-1} was lethal for *Atherina* larvae during 2 h since the beginning of the experiment. Concentration of 0.1 ng l^{-1} caused the significant decrease (more than 2-fold) heat production in larvae during 24 h. The number of protein fractions in larvae exposed to 1 ng l^{-1} PCB dropped. CAT and GR activities increased in fish larvae exposed to 50 ng l^{-1} PCB. SOD activity elevated upto 51% in fish incubated at concentration of 1500 ng l^{-1} during 1 h. The integral response of examined biomarkers in *Atherina* larvae exposed in various PCB concentrations may be applied as a good tool for evaluation of fish larvae health and ecological status of their environment.

Early life stages from a marine fish species, *Fundulus heteroclitus*, were exposed to sublethal doses of 3,30,4,40,5 pentachlorobiphenyl (PCB126) to evaluate its effects on ecologically relevant responses, growth and behavior. A few hours after fertilisation, eggs were treated topically with PCB126 which did not increase the mortality or malformation rates. Body length and spontaneous locomotor activity were altered only in larvae treated with the highest dose. Treatment with PCB126 caused a dose-responsive reduction in prey capture ability (rate of decline in the number of *Artemia*) and induction of EROD activity. In untreated developing larvae, prey capture ability and efficiency increased as post-hatching development progressed and EROD activity remained low. Four days post-hatching (dph), morphological changes (body length and malformations), spontaneous locomotor activity (active swimming speed, rate of travel, % inactivity), prey capture ability (*Artemia franciscana* nauplii) and whole body EROD activity were evaluated in larvae and it was shown that PCB126 did not increase the mortality or malformation rates. The pattern of behavioral responses observed in PCB126-exposed *Fundulus* larvae differed from that observed in less-developed larvae indicating that other mechanisms than retarded development were involved (Fortin et al., 2008).

Tetrachlorodibenzo-p-dioxin (TCDD) was frequently detected in environmental samples and caused several developmental toxic responses to teleost fish through the AHR/ARNT signaling pathway, particularly the yolk sac and pericardial edema. Zebrafish embryos were successfully applied to the toxicity investigation of TCDD to a relatively low concentration (50 pg ml^{-1}) for 72 h. Results showed that CYP1A level in TCDD-treated fish ultimately cropped down after an initial rise. The variety of CYP1A transcription paralleled with the TCDD exposure time, which reflected the bioaccumulation and metabolism of TCDD *in vivo*. Toxicity endpoints did not occur shortly after exposure, but took place from 72 hpf or later (Wu et al., 2008).

Zebrafish embryos were used to evaluate the developmental toxicity of individual hexabromocyclododecanes (HBCD) diastereoisomers (α-HBCD, β-HBCD and γ-HBCD). Four-hour post-fertilization (hpf) zebrafish embryos were exposed to different concentrations of HBCD diastereoisomers (0, 0.01, 0.1 and 1.0 mg/l) until 120 hpf. The results showed that exposure to HBCDs can affect the development of zebrafish embryos/larvae in a dose-dependent and diastereoselective manner. The diastereoisomersα-, β- and γ-HBCD had different effects on the development of zebrafish (*Danio rerio*) embryos such as hatching delay, growth inhibition, reduction of heart rate of larvae (96 hpf), significant increase in mortality and malformation rate. HBCD diastereoisomers could induce the generation of ROS and the activities of caspase-3 and caspase-9 in a dose-dependent manner. The results indicated that HBCD diastereoisomers could cause developmental toxicity to zebrafish embryos through inducing apoptosis by ROS formation. The results showed a good agreement confirming that the order of developmental toxicity of HBCD diastereoisomers in zebrafish is γ-HBCD >β-HBCD >α-HBCD (Du et al., 2012).

The next organic compound tetrabromobisphenol A (TBBPA) is a widely used brominated flame retardant that is persistent in the environment. TBBPA is microbiologically transformed in anaerobic environments to bisphenol A (BPA) and in aerobic environments to TBBPA dimethyl ether (TBBPA DME). The relative toxicity of TBBPA, BPA and TBBPA DME was determined using exposure to zebrafish embryos, with BPA and TBBPA DME exhibiting lower potency than TBBPA. TBBPA exposure resulted in 100% mortality at 3 (1.6 mg/L) and 1.5 μM (0.8 mg/L). While all three caused edema and hemorrhage, only TBBPA specifically caused decreased heart rate, edema of the trunk, and tail malformations. Matrix metalloproteinase (MMP) expression was measured due to the role of these enzymes in the remodeling of the extracellular matrix during tissue morphogenesis, wound healing and cell migration. MMP-2, -9 and -13 expression increased (2–8-fold) after TBBPA exposure followed by an increase in the degradation of collagen I and gelatin. TBBPA DME exposure resulted in only a slight increase (less than 2-fold) in MMP expression and did not significantly increase enzymatic activity. These data suggest that TBBPA is more potent than BPA or TBBPA DME and indicate that the trunk and tail phenotypes seen after TBBPA exposure could be due in part to alteration of proper MMP expression and activity (McCormick et al., 2010).

The investigators examined the transcriptional responses of perfluorooctanesulfonate (PFOS) accumulated in the marine medaka embryos at the early and late developmental stages of 4 and 10 dpf at different concentrations (1, 4, and 16 mg/L). PFOS accumulated in the embryos, and the embryonic burdens of PFOS at 10 dpf were markedly higher than those at 4 dpf. Thirteen genes involved in three important POPs-related receptor pathways, including ER, AHR and PPAR, were cloned and also investigated. The mRNA expression levels of ERα and ERγ were not significantly altered, but the estrogenic marker genes were down-regulated upon PFOS exposure at 4 dpf. Conversely, ERs and all related marker genes were significantly up-regulated at 10 dpf. The expressions of ARNT and cyp1a were both up-regulated at 4 dpf, while no obvious changes were detected at 10 dpf. The expressions of CYP19A and CYP19B were regulated by PFOS in a stage-specific manner. PFOS produced different effects on three isoforms of PPAR. PPARα and PPARβ were first inhibited at 4 dpf and were induced at 10 dpf. PFOS did not elicit a change in PPARγ expression at either stage. In conclusion, this study showed that PFOS has an estrogenic activity and endocrine-disruptive properties. Meanwhile, PFOS could elicit transcriptional responses on POPs-related pathways in a stage-specific manner (Fang et al., 2012).

The majority of studies characterizing the mechanisms of oil toxicity in fish embryos and larvae have focused largely on unrefined crude oil. At the same time polycyclic aromatic hydrocarbons (PAH) are hydrophobic environmental contaminants with petrogenic, biogenic, and pyrogenic sources. PAHs are organic compounds composed of two or more benzene rings fused together. This is the class of organic pollutants released in largest quantities in the environment. PAHs are generated mostly from the incomplete combustion of all kinds of carbon containing materials and from discharges of crude oils and other liquids and materials derived from fossil fuel. The PAHs appear as a highly diverse group of chemicals (Utvik, 1999). Alkylated PAH predominate in crude oils, are found in sediment downstream of pulp and paper mills, and can be more toxic than their non-alkylated homologues. The enzymatic metabolism of alkyl phenanthrenes generates ring and chain hydroxylated derivatives (Fallahtafti et al., 2012). Phenols were more toxic than benzylic alcohols, and some phenols were more than four times toxic than their non-hydroxylated counterpart to the early life stages of Japanese medaka (*Oryzias latipes*). Ring hydroxylation can increase PAH toxicity, and metabolism may enhance alkyl-PAH toxicity through the generation of such metabolites. The authors showed the toxicity of a suite of hydroxylated alkyl-PAH to the early life stages of fish, proposing an association between the preferential formation of *para*-quinones and enhanced toxicity (Fallahtafti et al., 2012).

Endpoints of planar halogenated aromatic hydrocarbon (pHAH) and polycyclic aromatic hydrocarbon (PAH) toxicity are mediated *via* activation of the aryl hydrocarbon receptor (AhR) followed by activation of the so called "AhR-battery" of genes including the cytochrome P450 1 (CYP1) isoforms. CYP-induced fluorescence allowed for the *in vivo* detection of CYP1 enzyme activity down to the cellular level as early as in the gastrulation stage of zebrafish. Basal and induced CYP1 activities were detected at all time points examined from 8 h post-fertilization to early adulthood and showed a highly dynamic spatio-temporal pattern throughout zebrafish development. Basal and induced EROD activity was prominent in tissues of the cardiovascular and digestive systems, the urinary tract, and parts of the brain as well as in the central portion of the eye and the otic vesicle during distinct stages of development. The differentiation between constitutive and induced spatio-temporal patterns of CYP1 activity even as early as the gastrula stage provide further insights into the endogenous role of CYP1 activity (Otte et al., 2010).

Benzo[a]pyrene (BaP) is a ubiquitous environmental polycyclic aromatic hydrocarbon (PAH) contaminant that is both a carcinogen and a developmental toxicant. Some of BaP's developmental toxicity may be mediated by effects on glycine N-methyltransferase (GNMT) which is a mediator in the methionine and folate cycles, and the homotetrameric form enzymatically transfers a methyl group from S-adenosylmethionine (SAM) to glycine forming S-adenosylhomocysteine (SAH) and sarcosine. SAM homeostasis, as regulated by GNMT, is critically involved in regulation of DNA methylation, and altered GNMT expression is associated with liver pathologies. The homodimeric form of GNMT has been suggested as the 4S PAH-binding protein. *Fundulus heteroclitus* embryos were exposed to waterborne BaP at 10 and 100 µg/L. Whole mount *in situ* hybridization showed that GNMT mRNA expression was increased by BaP in the liver region of 7, 10 and 14 dpf *F. heteroclitus* embryos. In contrast to mRNA induction, *in vivo* BaP exposure decreased GNMT enzyme activity in 4, 10 and 14 dpf embryos. BaP exposure altered GNMT expression, which may represent a new target pathway for BaP-mediated embryonic toxicities and DNA methylation changes (Fang et al., 2010).

High concentrations of PAHs in water cause hypoxia, which affects the embryotoxicity of PAHs. About 500 μgL^{-1} benzo[a]pyren (BaP) and 1–200 μgL^{-1} benzo[k]fluoranthene (BkF) interacted synergistically with hypoxia (7.5% oxygen, 35% of normoxia) to induce pericardial edema in developing zebrafish. Hypoxia protected from the embryotoxicity of pyren (PY) and had no effect on the toxicity of polychlorinated biphenyl-126. Interactions between hypoxia and BkF and PY were closely mimicked by morpholino knockdown of CYP1A, indicating a potential role for metabolism of these compounds in their toxicity. The investigators concluded that various PAHs may exhibit synergistic, antagonistic or additive toxicity with hypoxia and risk assessments may underestimate the threat of PAHs to fish early stage in contaminated sites (Fleming and Di Giulio, 2011).

Endocrine system and disruptors

The endocrine system consists of various glands which synthesize and secrete hormones that regulate development, growth, metabolism and reproduction. Environmental contaminants and toxicants including some of the alkylphenols, PAHs and PCBs can adversely affect endocrine system in wildlife including fish species. Biomarkers are currently used in the environmental risk assessment of endocrine-disrupting chemicals (EDCs) because many aquatic ecosystems receive significant inputs of natural and synthetic chemicals that act as endocrine disrupting chemicals, which constitute a threat to the reproductive health of fish populations. They include androgen and estrogen agonists and antagonists, aromatase inhibitors, and also thyroid disruptors (Hutchinson et al., 2006). Several studies indicate that exposure of fish to relatively high concentrations of endocrine disruptors (phthalate esters) during sensitive part of life cycle may interfere with gonad differentiation (Norman et al., 2007).

Thyroid hormones play an important role in regulation of development process, morphogenesis, growth, and behavior in fish. Some environmental pollutants have adverse effects on either development or function of the thyroid gland in fish. The investigation of rockfish (*Sebastiscus marmoratus*) embryos exposed to pyrene (PY) for 5 days at the concentrations of 0.5, 5, and 50 nmol/L showed that toxicant exposure decreased the expression of thyroid primordium markers, *Pax 2.1* and *Nk2.1a* and reduced the concentration of T_3, but not T_4. Thyroid receptor genes (*TRα* and *TRβ*) expression was down-regulated by Py. Py exposure impaired the expression of thyroid development related genes, *Fgfr2* and *Hoxa3a* expression, and altered the mRNA levels of thyroid function related genes, *Deio1*, *Ttr*, and *Tg*. In general, the authors noted that PY exposure inhibited thyroid development and influenced the function of thyroid system in rockfish embryos (He et al., 2012). The microcalorimetric study of thyroid-disrupting toxicants (bisphenol A and nonylphenol) observed that the chemicals modified metabolic rate of salmon embryos throughout the development which was connected with the energetic consequences by affecting regulation of metabolic rate (Penttinen et al., 2005).

Polybrominated diphenyl ethers (PBDEs) have the potential to also disturb the thyroid endocrine system. Bioconcentration and metabolism of BDE-209 were investigated in zebrafish embryos exposed at concentrations of 0, 0.08, 0.38 and 1.92 mg/L in water until 14 days post-fertilization (dpf). The alterations of both triiodothyronine (T_3) and thyroxine (T_4) levels, indicating thyroid endocrine disruption were noted. Gene transcription in the hypothalamic–pituitary–thyroid (HPT) axis study showed that the genes encoding corticotrophin-releasing hormone (*CRH*) and thyroid-stimulating hormone (*TSHβ*) were transcriptionally significantly up-regulated. Genes involved in thyroid development (*Pax8* and *Nkx2.1*), synthesis (sodium/iodide symporter, *NIS*, thyroglobulin, *TG*), mRNA for thyroninedeiodinase (*Dio1* and *Dio2*) and thyroid hormone receptors (*TRα* and *TRβ*) were also transcriptionally up-regulated. However, the genes encoding proteins involved in TH transport (transthyretin, *TTR*) and metabolism (uridinediphosphate-glucuronosyl-transferase, *UGT1ab*) were transcriptionally significantly down-regulated. Protein synthesis of TG was significantly up-regulated, while that of TTR was significantly reduced. The researchers suggested that the hypothalamic–pituitary–thyroid axis can be evaluated to determine thyroid endocrine disruption by BDE-209 in developing zebrafish larvae (Chen et al., 2012).

Other researchers performed waterborne exposures of 2,2′,4,4′-tetrabromodiphenyl ether (BDE-47), tetrabromobisphenol A (TBBPA) or bisphenol A (BPA) on zebrafish embryo–larvae and quantitatively measured the expression of genes belonging to the hypothalamic-pituitary-thyroid (HPT) axis to assess for adverse thyroid function. For analysis on the effects of BDE-47, TBBPA and BPA on the hypothalamic-pituitary-thyroid genes, zebrafish embryo–larvae were acutely exposed to lethal and sub-lethal concentrations of the chemicals. In larvae, BDE-47 was found to have significantly induced many genes such as thyroglobulin, thyroid peroxidase, thyroid receptors α and β, thyroid stimulating hormone, and transthyretin. TBBPA significantly induced only three genes: thyroid receptor α, thyroid stimulating hormone, and transthyretin, while BPA only induced thyroid stimulating hormone. In embryos, BDE-47 significantly induced the sodium iodide symporter and thyroid stimulating hormone. TBBPA significantly induced thyroid receptor α and thyroid stimulating hormone, while BPA did not significantly induce any of the genes. Most genes were only induced at the 75% 96 h-LC50 or 96 h-EC50 value; however, thyroid peroxidase and thyroid stimulating hormone demonstrated up-regulation in a level as little as the 10% 96 h-LC50 value. The authors proposed that the obtained results could be useful for elucidating the toxicological mechanism of brominated flame retardants, assessing appropriate safety levels in the environment for these compounds, as well as serve as a reference for other man-made contaminants (Chan and Chan, 2012).

Plasma vitellogenin production in adult fish male exposed to various kinds of EDCs were documented (Pait and Nelson, 2003), while the information of the response of early development stages is scare. Exposure of Atlantic cod (*Gadus morhua*) to different levels of North Sea produced water (PW) and 17β-oestradiol (E_2), a natural oestrogen, from egg to fry stage (90 days) showed the changes in protein expression following E_2 exposure to changes induced by PW treatment. Many of the protein changes in whole cod fry (approximately 80 days post-hatching, dph) occurred at low levels (0.01% and 0.1% PW) of exposure, indicating putative biological responses at lower levels. The biomarker candidates could be effective diagnostic tools in monitoring exposure and effects of discharges from the industries and sewage containing EDCs chemicals (Bohne-Kjersem et al., 2010). The investigators observed intersex (1/80) of juvenile chub *Leuciscus cephalus*, exposed to mixture of 11-ketotestosterone and 17 β-estradiol. The combination of the hormones significantly increased whole body VTG level (Zlabek et al., 2009).

Biomarkers of exposure to these stressors, cytochrome P4501a (CYP1A), estrogen receptor alpha (ERα), brain cytochrome P450 aromatase (CYP19a2 or AromB), and hypoxia inducible factor 1 alpha (Hif-1α) mRNA expression were examined using qRT-PCR, simultaneously in embryos of two species, the Atlantic killifish, *Fundulus heteroclitus*, and the zebrafish *Danio rerio*. Embryos of both species were exposed to the model CYP1A inducer β-naphthoflavone (BNF) or 17-β estradiol (E2) under either normoxic or hypoxic (5% oxygen atmosphere) conditions and harvested prior to hatch at 9 days post fertilization (dpf) for the killifish, and 48 h post fertilization (hpf) for the zebrafish. BNF significantly induced CYP1A expression in embryos of both species. However, killifish embryos were more responsive (700-fold > control) than zebrafish embryos (7-100-fold > control). AromB was also significantly influenced by treatment, but to a lesser extent, while mean expression levels increased by less than two-fold over control values in response to E2, and in one case up-regulated by BNF. ERα and Hif-1α were constitutively expressed in embryos of both species, but expression was unaffected by exposure to either BNF or E2. Hypoxic conditions down regulated AromB expression strongly in killifish but not in zebrafish embryos. The results support the use of CYP1A expression as a biomarker of AhR agonists in fish embryos, and indicate that AromB may be more responsive than ERα to estrogenic chemicals at this stage of development. Finally killifish embryos are generally more sensitive than zebrafish embryos at this stage of development supporting their use in environmental assessments (McElroy et al., 2012).

Pesticides

Pesticides entering into marine ecosystem throw river discharge and domestic sewage which resulted in great stress in marine organisms, including early developmental stages of fish. Cuprocsat ($CuSO_4$ $3Cu(OH)_2$ $1/2H_2O$) and cyfose (phosphorus containing organic pesticide) are widely used in agriculture of the southern Ukraine region (Rudneva and Shaida, 2012). We observed the toxic response of fish larvae and we proposed the metabolic rate of fish as biomarker of toxicity (Fig. 4.13, 4.14).

Heat output of *A. mochon pontica* larvae exposed in cuprocsat in all tested concentrations was more than 2-fold lower as compared to the values of the intact animals ($p < 0.01$). At the same time the variables between the heat production of the larvae exposed to the concentrations of 0.625 and 1.25 mg·l⁻¹ were insignificant while the metabolic rate of the animals exposed to the highest concentration was significantly lower ($p < 0.01$) than the control and the other examined groups.

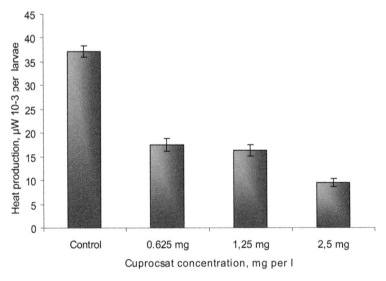

Fig. 4.13. Metabolic rate *A. mochon pontica* larvae exposed to cuprocsat (n = 3, mean+ SEM). Time of the monitoring of the Monitor of biological activity was 24h at the temperature +20°C.

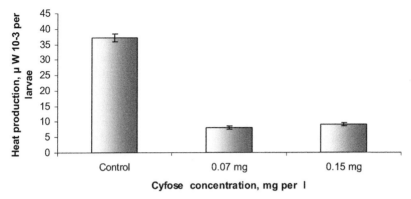

Fig. 4.14. Metabolic rate *A. mochon pontica* larvae exposed to cyfose (n = 3, mean+ SEM). Time of the monitoring in of the Monitor of biological activity was 24h at the temperature +20°C.

Heat output of *A. mochon pontica* larvae exposed to both concentrations of cyfose decreased significantly (p < 0.01) as compared to the control animals. However, no differences were shown between the values of heat dissipation of both experimental groups.

The comparative study of the pesticide effects on the metabolic rate demonstrated the similar decrease of the heat dissipation in fish larvae. In the case of organic cyfose we could propose that the mechanism of its toxicity and the influence on heat production was comparable with the effects of PCB and other xenobiotics of the similar origin (Rudneva and Shaida, 2006; Rudneva et al., 2004a,b).

In the case of cuprocsat, cooper—containing pesticide, mechanism of heat production modification in treated fish larvae could be differed. Metals are highly distributed pollutants in aquatic ecosystems and exposure to them disturbs metabolic processes in fish and invertebrates. The negative effects are characterized by the changes in respiratory function. Copper ions damage gill epithelium, oxygen concentration in tissues decreases and metabolic rate of fish changes. The deficiency in oxygen and energy decrease to the resistance or repairation abilities of the organism to toxicant-induced damage (Hopkins et al., 2003; Pane et al., 2003).

It's well-documented that exposure to toxicants led to the negative effects on aquatic animals including fish and invertebrates which associated with the damage of respiratory, swimming behavior, histopathology, accumulation in body tissues and damage of biochemical pathways. In our previous investigations we demonstrated the modification of protein and lipid metabolism, elevation of the lipid peroxidation level and antioxidant enzyme activities in early developmental stages of fish and invertebrates (Rudneva et al., 2005; Rudneva and Shaida, 2006; Zalevskaya et al., 2004). Several investigators documented the increase of energy requirements to resist or repair toxicant-induced damage in aquatic animals. In contrast, some chemicals are known to decrease metabolic rate or they don't change it (Hopkins et al., 2003). The reasons of the metabolism modification in aquatic animals exposed to toxicants could be explained with the hypometabolism likely stems from damage gill epithelium, neutoxicity or hypoactivity and these three factors can also ultimately affect growth, survival and reproduction (Hopkins et al., 2003). At the same time despite the high concentrations of Hg accumulated by mosquitofish in the experimental conditions no difference in metabolic rate among individual fish from three populations were found (Hopkins et al., 2003). In contrast the acute exposure to dissolved inorganic Hg increased metabolic rate of mosquitofish from the same source population (Tatara et al., 1999). The authors suggested that the chronic environmental contamination led to the general resistance of fish to Hg and subsequent attenuation of effects on metabolic rate (Hopkins et al., 2003). In our study cuprocsat (cooper containing pesticide) was toxic for fish larvae led to decrease of heat production. It should be result the metabolism alterations and the enhance of reactive oxygen species (ROS) production which we noted previously (Rudneva, 1998; Zalevskaya et al., 2004). ROS are damaged by a lot of biological substances such as DNA, lipids and enzymes involving in the major metabolic pathways. High production of ROS impacted negatively on physiological status of the organism and disturbed its metabolism including energy generation and utilization balance.

We also studied the effect of the fungicide cuprocsat ($CuSO_4 \, 3Cu(OH)_2 \, 1/2H_2O$) at the concentrations of 0.15, 0.312, 0,625, 1.25 and 2.5 mg L^{-1} on *Atherina hepsetus* larvae due 4 days of exposure (Rudneva et al., 2004a). The results obtained showed the variations of antioxidant enzyme activities which were depended on pesticide concentration and exposure time (Fig. 4.15). In all experimental groups of larvae CAT activity was higher as compared to the control during the period of exposure, especially in low and medium concentrations of cuprocsat. SOD activity varied lower, it increased in all tested concentrations at day 2. PER activity tended to enhance at low and medium concentrations of pesticide, and decreased in high concentrations. GR activity was higher than the values of intact larvae at days 1 and 2 of exposure at low and medium concentrations (with the exception of 1 day at 0.625 mg L^{-1}), then tended to decline at high concentrations at the end of exposure. GST activity was greater in all experimental groups at day 2 especially in high concentrations. Therefore, the most sensitive enzymes to cuprocsat toxicity were CAT (increase of activity to 1500% as compared to the control) and SOD (increase of activity to 600% as compared to the control) at the 1 day of exposure at the lowest concentration (0.15 mg L^{-1}). The response of antioxidant enzyme activities

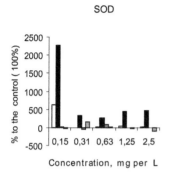

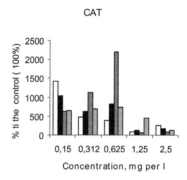

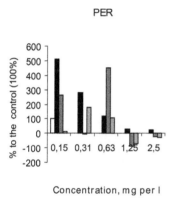

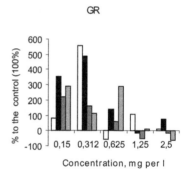

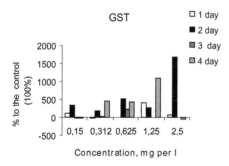

Fig. 4.15. Antioxidant enzyme activities of the *Atherina* larvae exposed to different concentrations of fungicide cuprocsat during four days of exposure. CAT-catalase, SOD- superoxide dismutase, PER-peroxidase, GR-glutathione reductase, GST-glutathione-S-transferase activities estimated relatively to the control as 100%.

was not uniform and depended on pesticide concentration and time of exposure. However, the correlation between antioxidant enzyme response, time of exposure and pesticide concentration were non-liner.

In addition, we noted the enhancement of lipid peroxidation production (200–1000% as compared to the control) especially at 3 and 4 days of exposure (Rudneva et al., 2004a). On contrary, the changes of low molecular weight antioxidants (vitamin A and carotenoids) levels were not uniform. Vitamin A concentration decreased as compared to the control to 15–75% while carotenoids values varied unclearly (Fig. 4.16).

Therfore, pesticides induce oxidative stress in fish early developmental stages and the response of biomarkers (lipid peroxidation and antioxidants levels) depend on kind of pesticides, concentrations and duration of exposure.

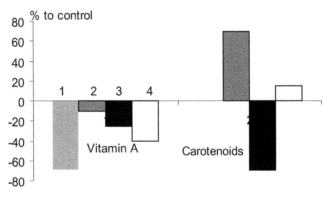

Fig. 4.16. Trends of vitamin A and carotenoids concentration in *Atherina* larvae exposed to fungicide cuprocsat (0.625 mg L^{-1}) to control as 100%. 1,2,3,4–days of exposure.

Detergents and PCP products

Information about marine organisms exposed to detergents is scarce and studies about internal changes and biomarker response in early-life stages of fish due to exposure to surfactants have been limited. Generally, investigators reported the effects of detergents on adult fish. For example, the toxicity of sublethal concentrations of effluents from a soap and detergent industry were studied on African catfish *Clarias gariepinus* using a renewable static bioassay with the association of heavy metals accumulation. The bioconcentration of metals in gut of test organisms was significantly higher than in muscle ($p < 0.05$). Swimming and feeding activities of fish reduced drastically and they became very weak since they could no longer feed well. Colour changed from black to pale black with mucus covering the entire body in response to the toxic effect of the detergent effluents.

This might have resulted due to the excretion. The results of this study revealed that fish can accumulate heavy metals from detergents-polluted environment that causes damage to their physiological status and the loss of natural population size (Ayandiran et al., 2009).

Personal care products (PCPs) such as shampoos, fragrances, cosmetics, toothpaste and soap, including biocide triclosan (TCS) are commonly used in personal care, acrylic, plastic, and textiles products. TCS showed acute toxicity for fish embryos and larvae (96 hLC50 = 0.42 mg/l) and delayed hatching. Embryotoxicity characterized the following: delay on the otolith formation and eye and body pigmentation, spine malformations, pericardial oedema and undersize. ChE and LDH activity were increased in larvae exposed to 0.25 mg/l, and GST activity was increased in larvae exposed to 0.25 and 0.35 mg/l. Despite the fact that similar 96 h LC50 values have been found for *Dentex rerio* embryos and adults (0.42 and 0.34 mg/l, respectively), the embryo assay was much more informative, showing important effects at several levels, including teratogenic response, hatching delay and alteration of biomarker levels. Analysis of biomarkers levels on larvae exposed for 96 h to triclosan showed important differences on the activities at concentrations that had not induced any effect at survival or embryo development level (e.g., 0.25 mg/l), suggesting that enzymatic levels are very sensitive parameters for embryo/larvae TCS toxicity assessment. TCS does not seem to be genotoxic for adult fish or to interfere with biomarker levels at the concentrations tested. TCS has deleterious effects on zebrafish adults and during early stages, including embryotoxicity, hatching delay and alterations of biomarker levels. The results obtained suggested high toxicity of TCS with adverse effects on survival, embryonic development hatching and enzyme activities. The GST, ChE, LDH activity were the most sensitive endpoints and biomarkers for TCS toxicity assessment (Oliveira et al., 2009).

There are several studies of the toxic effects of linear alkylbenzene sulphonate (LAS), the most widely used anionic surfactant in household and cleaning products. In fish, surfactants caused damage in the gills, skin and pharynx in fish. They may penetrate into the organism and exert adverse effects on the internal functions. The result of damage is generally manifested by loss of orientation, tendency to swallow air, lethargy or death. Neonate (< 24h) larvae of the sea bream, *Sparus aurata*, were exposed for 72h to sublethal concentrations (0.1–1.0 mg·L^{-1}) of LAS. The response and degree of the observed effects of LAS exposure on sea bream larvae appeared to depend on both surfactant concentration and exposure time. In LAS-exposed larvae only a weak increase of CYP1A was observed in the hepatic vascular system at LAS concentrations ranging between 0.4 and 0.5 mg·L^{-1}. No different LAS concentrations could be established. Surface active compounds as surfactants are composed of a hydrophilic part and a lipophilic alkyl chain. The higher the number of carbon atoms in the alkyl-chain, the more lipophilic is the molecule. It is possible that due to the presence of a polar group, LAS may not be an adequate CYP1A inductor (Hampel et al., 2004).

In fingerlings of *Cyprinus carpio* exposed to sublethal concentrations of LAS, alterations in the levels of glycogen, lactic acid, sialic acid, and acid and alkaline phosphatases in the gills, liver and kidney at concentrations of 0.005 mg·L^{-1} were found (Misra et al., 1991). Nevertheless, as early life stages do not have completely developed organs, sublethal effects of LAS exposure to early life stages may manifest themselves in a different way to that mentioned by these authors.

Effluents

Domestic sewage contains the mixture of various kinds of chemicals which directly and indirectly affect the fish health. Stress response in fish *Paracentrotus lividus* embryo-larval exposed in sewage-influenced seawater was shown. The authors suggested that the use of biomarkers was very important to predict ecological effects from total pollution of sewage (Sánchez-Marín et al., 2010).

The study of response of *Lepadogaster lepadogaster lepadogaster* embryos exposed to domestic sewage in different concentrations indicated that the time of hatching process was increased and heat production was also greater in treated hatching larvae as compared to the impact individuals (Kuzminova et al., 2004). In addition, the researcher showed that SOD, CAT, PER and GR activities were higher in the *A. mochon pontica* larvae incubated in domestic sewage at the concentrations ranged from 1:1000 to 1:10 (sewage/marine water) as compared in control while the heat production was decreased with the increasing of sewage level (Kuzminova, 2003).

Atlantic cod (*Gadus morhua*) were exposed to different levels of produced water (PW) which contains numerous toxic compounds of dispersed oil, metals, alkylphenols (APs), polycyclic aromatic hydrocarbons (PAHs) and 17β-oestradiol (E$_2$), a natural oestrogen, from egg to fry stage (90 days). Changes in the proteome in response to exposure in whole cod fry (approximately 80 days post-hatching, dph) occurred at low levels (0.01% and 0.1% PW) of exposure, indicating putative biological responses. The biomarkers may, following validation, prove effective as diagnostic tools in monitoring exposure and effects of discharges from the petroleum industry offshore, aiding future environmental risk analysis and risk management (Bohne-Kjersem et al., 2010).

Hence, examined biomarkers in early developmental stages of fish responded on various kinds of sewage impact in laboratory conditions. Laboratory studies dealing with multiple toxicants interactions should be performed to enable a better understanding of mechanisms of their toxicity in the aquatic environment. However, aquatic contamination in field ecosystems involves various chemicals that interact with one another and lead to synergistic or antagonistic effects. For that reason, studies on the effects of contaminants mixtures in field conditions are required.

4.2 Field Studies

Results of several studies show that embryo malformation rate and biomarker response can be used as indicators of environmental pollution and water quality assessment. For instance, the authors noted the highest mean levels of embryo deformity (20–30%) in the regions of poor water quality near Xiamen Harbour (highly urbanized and industrialized region of China) and these decreased towards open waters where abnormalities approached background levels. At the same time activities of EROD and GST in adult fish indicated no clear pattern. Antioxidant biomarkers (GP, CAT, SOD and reduced glutathione) in fish liver from tested areas suggested that exposure to xenobiotics appeared to be the lowest in reference sites and the highest in polluted areas. Inhibition of AChE in fish muscle was also observed in high polluted site (Klumpp et al., 2002).

The effects of Chilean pulp mill effluent extracts (untreated, primary and secondary treated pulp mill effluents), along with steroid standards (testosterone and 17β-estradiol) and a wood extractive standard (beta-sitosterol) on developing post-fertilized fish embryos were studied on a cold freshwater species, rainbow trout (*Oncorhynchusmykiss*), and two warm freshwater species American flagfish (*Jordanellafloridae*) and Japanese medaka (*Oryziaslatipes*). Embryotoxicity results demonstrated the delay in time to hatch and decreased hatchability but no significant egg and larvae mortality was observed in the pulp mill extract exposed embryos. In contrast, significant early hatching and increased hatchability were observed in beta-sitosterol exposed embryos, along with high mortality of testosterone exposed embryos across species. Teratogenic responses were observed in medaka embryos in all treatments. Abnormalities were detected at development stages 19–20 (2–4 somite stages) and included optical deformities (micro-opthalmia, 1 or 2 eyes) and lack of development of brain and heart. Additionally, phenotypic sex identification of surviving offspring found female-biased sex-ratios in all treatments except testosterone across species. Overall, the study indicated that Chilean pulp and paper mill extractives caused embryotoxicity (post-fertilized embryos) across species irrespective of the effluent treatment. The effects were mainly associated with delayed time to hatch, decreased hatchability, and species-specific teratogenesis (Orrego et al., 2011).

Sardine larvae (*Sardina pilchardus*) were sampled from four sites along a transect out of the Bilbao estuary and a fifth (reference) site further along the north coast of Spain. The authors detected that EROD activity was the lowest at a site close to the Bilbao estuary, and increased along the transect until it reached levels recorded from samples for a reference site. CAT and SOD activities were the highest at the inshore site of the transect decreasing offshore before increasing again at the site farthest from the coastline. The researchers noted that factors other than PAH and PCB loading were affecting CYP1A catalytic activity in sardine larvae because they did not find strong correlation between PAH and PCB levels in zooplankton and enzymatic activities of tested biomarkers (Peters et al., 1994).

The antioxidant enzymes SOD and CAT study in the subcellular fraction of the clupeid larvae *S. sprattus*, sampled from the 11 sites of the southern North Sea demonstrated that the highest enzymatic activities were indicated at stations near the outflow of the Elbe and Weser rivers which were associated with the greatest levels of PCB and DDE in zooplankton. SOD and CAT activities were indicated to decrease along two transects running north and north-west from the discharge of two rivers. The authors suggested that in high polluted sites the resistance against oxidative stress increased in *S. sprattus* larvae (Peters et al., 2001).

As we noted above (see 4.1) mercury is recognized as a potent neurotoxicant and affects fish health and behavior. Adults of European sea bass (*Dicentrarchus labrax*) captured in an estuarine area affected by chlor-alkalin industry discharges (Laranjo Basin, Ria de Aveiro, Portugal) was tested in warm and cold periods. In the warm period, brain of fish from mercury contaminated sites exhibited ambivalent antioxidant responses. Higher GR activity and lower CAT activity respectively, were regarded, as possible signs of protective adaptation and increased susceptibility to oxidative stress challenge. Though the risk of an overwhelming ROS production cannot be excluded, brain appeared to possess compensatory mechanisms and was able to avoid lipid peroxidative damage. The warm period was the most critical for the appearance of oxidative damage as no inter-site alterations on oxidative stress endpoints were detected in the cold period. Since seasonal differences were found in oxidative stress responses and not in mercury bioaccumulation, environmental factors affected the former more than the latter (Mieiro et al., 2011).

Domestic sewage containing many toxic xenobiotics (heavy metals, organic compounds, pesticides, PCB, DDT and their metabolites, etc.) impact coastal ecosystems and leads to very negative biological events both for marine ecosystem and aquatic organisms especially in early development stages. The coastal areas are the spawning sites for many kinds of marine invertebrates and fish and high levels of pollution affect the eggs and larvae resulting in great malformation rate and abnormalities of development and growth. Domestic sewage and the effluents from sewage treatment plants strongly influence the water quality and flows into various water bodies which are their recipients. These effluents play a crucial role in maintaining the aquatic communities of these ecosystems, particularly in the absence of natural flow resulting from climate constraints or intensive water use. The results of ecological effects of these effluents were generally studied in adult fish and the information of early-life stages is very scarce. For instance, determination with non-lethal biomarkers in *Barbus meridionalis* demonstrated the utility of hematological parameters. On contrary in fish at reference sites, a decrease in hematocrit and hemoglobin, neutrophilia, lymphopenia, monocytosis, a rise in the nucleo-cytoplasmatic ratio of erythrocytes and an increase in the frequency of abnormal, immature and senescent erythrocytes were detected. Many hematological parameters correlated significantly with the environmental parameters measured. In addition to these hematological alterations, histopathological examination also revealed damage in fish livers but no impact was detected by the regional index of biotic integrity using fish as bioindicators. Battery of hematological parameters as biomarkers on a freshwater fish in a Mediterranean stream was a good tool for the evaluation of fish health in a region in which more than 50% of native fish species are classified as endangered or vulnerable (Maceda-Veiga et al., 2010).

Previously we described the high anthropogenic impact in Sevastopol Bay (Black Sea) and its negative consequences on fish health (Rudneva, 2011; Rudneva and Petzold-Bradley, 2001). The interspecies variations of antioxidant enzyme activities in fish tissues reflected the specific adaptations to the oxidative stress and protective mechanisms against oxidative damage. Our studies revealed these observations both in experimental and field conditions. The results of experimental studies have been described previously (see 4.1) and the studies of fish larvae inhabiting Sevastopol bays with different levels of anthropogenic impact are present below. Sevastopol bays characterized different levels of anthropogenic impact (Fig. 4.17). The bays are recipients of domestic, industrial and maritime transport sewage which enter into the sea and impact the ecosystem and biota (Rudneva and Petzold-Bradley, 2001; Rudneva, 2011).

Atherina hepsetus larvae (Fig. 4.18) were used as biomonitors in the examination of anthropognic impact (Rudneva and Zalevskaya, 2004). These are highly distributed fish species in Black Sea in various marine locations. *Atherina* larvae (size 8.0–8.2 mm) were collected in five Sevastopol bays with different level of anthropogenic impact: Sevastopolskaya, Artilleriyskaya, Streletskaya, Martynova and Kazach'ya (reference site). The results demonstrated that the anthropogenic pollution caused the responses of biomarkers in the larvae inhabiting the examined locations (Table 4.4).

No significant differences were observed in CAT activity in larvae from examined sites. SOD activity was significantly higher ($p < 0.01$) in fish from Sevastopolskaya and Artilleriyskaya bays than those in fish from Martynova

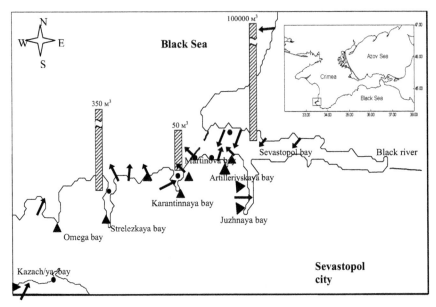

—fish catch sites, ▲—rain sewage, ↓—domestic sewage, the column —volume of domestic sewage, m³

Fig. 4.17. Sampling sites in Sevastopol bays (Black Sea, Ukraine).

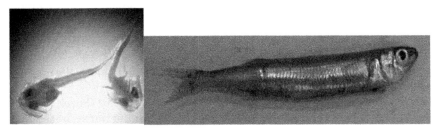

Fig. 4.18. *Atherina hepsetus* adult and its larvae.

Color image of this figure appears in the color plate section at the end of the book.

Table 4.4. Antioxidant enzyme activities (mg protein^{-1} min^{-1}, mean ± SEM, n = 3–5) in *Atherina* larvae from Sevastopol bays (Rudneva and Zalevskaya, 2004).

Bays	SOD, arbitrary units	CAT, mg H$_2$O$_2$	GR, nmol NADPH
Sevastopolskaya	5.408 ± 0.329** ***	0.026 ± 0.001	1.299 ± 0.094* ** ***
Artilleriyskaya	4.527 ± 0.972** ***	0.024 ± 0.001	1.461 ± 0.071* ** ***
Streletskaya	3.479 ± 0.847**	0.026 ± 0.003	0.723 ± 0.037 ***
Martynova	0.877 ± 0.247	0.026 ± 0.001	0.619 ± 0.054 ***
Kazach'ya	1.333 ± 0.646	0.023 ± 0.003	0.193 ± 0.038

*—differences are significant to the values of fish from Streletskaya Bay, **—from Martynova Bay, ***—from Kazach'ya Bay.

and Kazach'ya. SOD activity in larvae from Streletskaya Bay was also greater (p < 0.01) as compared to the values of the larvae from Martynova Bay. GR activity was the similar in the animals from Sevastopolskaya and Artilleriyskaya bays and in fish from Streletskaya and Martynova. The greatest SOD activity was indicated in fish from the most polluted areas (Sevastopolskaya and Artilleriyskaya bays) and the least values were shown in fish from Martynova Bay.

The significant differences of values of SOD and GR activity in the *Atherina* larvae demonstrated the polluted response of the fish to anthropogenic impact and the induction of defense antioxidant system in larvae collected in Sevastopolskaya, Artilleriyskaya and Strelatzskaya bays as compared to the larvae from Martynova and Kazach'ya. The changes in enzyme activities were assembled with the modifications of population, reproductive, morphological and physiological parameters of fish from highly contaminated areas and these effects were documented by the other investigators also (Livingston et al., 1995).

The obtained results have been shown that the examined biomarkers of fish larvae (with the exception of CAT) correlated with the anthropogenic pollution in the bays. High correlation coefficients between pollution level (effluents volume entering the bays) and SOD and GR activity were observed (r = 0.86 and r = 0.77 respectively). Low molecular weight antioxidants in fish larvae collected in five Sevastopol bays are present in Table 4.5. In this case the most polluted site is Juzhnaya Bay and the reference area is Omega Bay.

Carotenoids and vitamin A levels were significantly lower in larvae collected from polluted bays as compared to the reference site. High correlation coefficients were noted between the pollution level in bays and low molecular weight antioxidants which was ranged from r = 0.83 to r = 0.93 and the equation of this regression was the following Y = A + B/X + C/X². Therefore, the decrease of antioxidants concentration in fish larvae from polluted sites indicated their involvement into detoxification processes and defense mechanisms against oxidative stress. On the other hand the parameters of oxidative stress may be used as good tools for the

Table 4.5. Low molecular weight antioxidants level in larvae *A. mochon pontica* from Sevastopol bays (mean ± SEM, n = 3–5).

Bays	Carotenoids, µg/100 g lipids	Vitamin A mg/g lipids
Omega	31.24 ± 3.21	46.17 ± 2.64
Martynova	14.35 ± 2.31	24.68 ± 1.35
Sevastopolskaya	14.36 ± 1.98	28.74 ± 1.58
Streletskaya	15.50 ± 3.42	22.30 ± 2.04
Juzhnaya	16.80 ± 1.60	19.56 ± 1.67

evaluation of fish larvae *A. mochon pontica* status and water quality of their habitat. Lipid peroxidation level in fish from tested polluted locations is present in Table 4.6.

The data obtained showed that the lipid peroxidation compounds increased progressively in fish larvae collected in polluted sites as compared to the reference area. On the other hand, the values of fish from Martynova and Sevastopolskaya bays were the similar which could be explained by the similar level of pollution in both these locations. Generally, increase in the lipid peroxidation products in fish larvae from polluted sites reflecting the ROS induction resulted in high concentrations of pollutants in the environment. High correlation between lipid peroxidation values and pollution level in *A. mochon pontica* from tested sites was indicated ($r = 0.94$ for dien conjugates, $r = 0,84$ for ketodiens, $r = 0.96$ for hydroperoxides, and $r = 0.71$ for TBA-reactive compounds). The regression was not liner and the equation was the following: $Y = X/AX + B$. Hence, lipid peroxidation parameters could be used as good tools for evaluation of larvae health and water quality.

Electrophoretic composition of the whole protein extracts of fish larvae from polluted and non-polluted sites was also different (Rudneva and Zalevskaya, 2004). The findings demonstrated that protein composition of larvae *A. hepsetus* from high polluted Sevastopol bays (Sevastopolskaya and Artilleryiskaya, see Fig. 4.17) characterized high heterogeneity as compared to the electrophoretic spectra (EF spectra) of fish from reference area. In EF-spectra of larvae from high contaminated sites 5–6 protein fractions were detected while in fish from non-polluted location their number dropped to 2-fold. We proposed that high heterogeneity of EF-spectra of larvae from polluted sites could be the result of protein modification by the chemicals contained in the sewage. Xenobiotics may bind with proteins or (and) change of gene structure coding the proteins. In both cases the alteration of proteins was indicated in electrophoregramms.

Thus *Atherina* larvae could be applied as good test-organism in monitoring studies of Sevastopol coastal waters and their responses could characterize the pollution level of locations together with another biomarker

Table 4.6. Lipid peroxidation level in larvae of *A. mochon pontica* from Sevastopol bays (mg^{-1} lipids, mean ± SEM, n = 3–5).

Bays	Dien conjugates, μmol, x10^{-3}	Ketodiens, E270/215 x 10^{-2}	Hydroperoxides μg I$_2$	TBA-reactive compounds, nmol
Omega	4.8 ± 0.7	2.4 ± 0.9	35.2 ± 2.4	23.1 ± 2.3
Martynova	7.4 ± 1.2	11.8 ± 0.6	52.8 ± 2.5	33.5 ± 1.7
Sevastopolskaya	7.8 ± 0.9	8.1 ± 1.1	55.0 ± 2.5	29.6 ± 2.4
Streletskaya	11.3 ± 0.5	14.4 ± 0.9	58.4 ± 2.6	38.3 ± 1.4
Juzhnaya	12.4 ± 1.0	15.1 ± 1.2	61.2 ± 3.4	51.1 ± 3.2

and bioindicator (Karakoc et al., 1998). At the same time the larvae may be collected in the marine areas only during the warm period (spring-summer) but for ecotoxicological purposes it is important to use the biomonitor which could be collected in all seasons of the year (Goksoyr et al., 1996).

We also examined the response of metabolic rate of *A. hepsetus* collected in highly polluted and non-polluted marine areas. The 1-day larvae (length 8.0–8.2 mm with small yolk sac) were collected in polluted area (maritime area in Sebvastopol region) and in non-polluted location near Chersones protection territory (Rudneva and Shaida, 2012). The heat production trends of the larvae from both locations are present in Fig. 4.19.

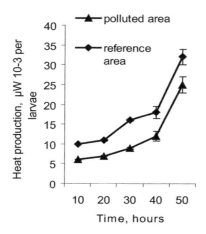

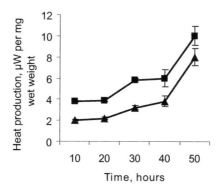

Fig. 4.19. Metabolic rate of *A. hepsetus* larvae caught in polluted and non-polluted locations (n = 5, mean ± SEM). Time of the monitoring in Monitor of biological activity was 50 h at the temperature +20°C.

The obtained results demonstrated the similar trends of heat production in both larvae during 50 h of monitoring in TAM. The curve trends reflect the period of stress and adaptation during the incubation in TAM. At the same time at the period of normal physiological state (20–40 h) heat output of the larvae from polluted site was significantly lower ($p < 0.01$) than the values of the fish from reference site. This fact could be explained due to the disturbance of the processes of energy generation and utilization in fish from the polluted habitats. However, throughout the period of experiment the oxygen concentration and food in yolk sac decreased and the differences between metabolic rate of two groups of larvae were also dropped. Thus we could conclude that the mixture of environmental pollutants containing in domestic sewage is a strong modifying factor of the metabolic rate in the early development stages of fish. It agrees with the observations of some researchers who noted that the thyroid-disrupting toxicants (bisphenol A and nonylphenol) were modified metabolic rate of salmon embryos throughout the development which was connected with the energetic consequences by affecting regulation of metabolic rate (Penttinen et al., 2005). Similar trend was shown in the fish larvae *Lepadogaster lepadogaster lepadogaster* (Bonnaterre) exposed to various concentrations of domestic sewage (1, 10 and 100 ml l^{-1}), containing the mixture of toxicants including the estrogens. The heat output of exposed larvae increased in series, but the heat production of developing embryos of this fish species was not sensitive to the sewage impact (Kuzminova, 2003).

Chronic impact of environmental pollution on fish energetic metabolism was also documented (Hopkins et al., 2003; Pane et al., 2003; Rudneva, 1998; Rudneva et al., 1998). Our findings demonstrated that the fish larvae inhabited high polluted marine sites impact many toxicants which caused energetic metabolism disturbance which was documented by other investigators also (Hopkins et al., 2003). Thus many environmental factors biotic, abiotic and anthropogenic may modify fish metabolism and the application of microcalorimetry with other analytical methods should be helpful for the explanations of the mechanisms of the developmental changes during embryogenesis, the influence of xenobiotics disturbance and interactions between the developing embryo and environment. Metabolic rate of *A. hepsetus* larvae inhabited highly polluted area was significantly lower as compared to the heat production of the larvae from the non-polluted area.

Heat output of aquatic organisms is very sensitive biomarker for determination of water toxicity and thus it could be applied as a good tool for the evaluation of water quality and has taken an effective role in solving many environmental problems such as the interactions between environmental factors and fish development, toxic effects and fish protection. Sensitive endpoints should be used in risk assessment because already low environmental concentrations of toxicants may cause damage of the organisms especially in early developmental stages of fish.

Therefore, the experimental conditions in field studies vary with season, physical and chemical properties of water which play an important role in pollutants toxicity. Organic substances in water interact with each other and with metals and modify toxicity effects. On the other hand, the senstivity of the organism to toxic effects depends on fish species, sex, size, maturation and age. Another factor that can influence a representative sampling is fish migration. Fish can be used as bioindicators of pollution in the aquatic environment by studying biomarkers; however, the specific forms of biomarkers and mechanisms of their action still need to be investigated.

The sensitivity of fish embryos, larvae and fry to pollutants is stage-dependent and not related to differences in uptake rate or bioavailability. Often this sensitivity is a function of the organism's ability to metabolize the toxicants or activate stress responses and repair mechanisms. In developing embryos cells are continually changing their molecular and physiological characteristics as they differentiate, and that early-life stage toxicity assays have been aimed primarily at search good biomarkers assessing negative events in developing embryos and larvae.

There are a number of toxins that are known to be embryotoxic or teratogenic causing embryo and larvae mortality and unsuccessful hatching and development. However, application of biomarkers reflecting molecular and cellular events occurring during embryonic development should be integrated into toxicological assessment programes.

References

Ayandiran, T.A., O.O. Fawole, S.O. Adewoye and M.A. ogundiran. 2009. Bioconcentration of metals in the body muscle and gut of *Clarias gariepinus* exposed to sublethal concentrations of soap and detergent effluent J. Cell and Animal Biology. 3(8): 113–118.

Barata, C., I. Lekumberri, M. Vila-Escale, N. Prat and C. Porte. 2005. Trace metal concentration, antioxidant enzyme activities and susceptibility to oxidative stress in the tricoptera larvae *Hydropsyche exocellata* from the Llobregat river basin (NE Spain). Aquatic Toxicol. 1: 3–19.

Bellas, J., R. Beiras, J.C. Marino-Balsa and N. Fernandez. 2005. Toxicity of organic compounds to marine invertebrate embryos and larvae: a comparison between sea urchin embryogenesis bioassay and alternative test species. Ecotoxicology. 14: 337–353.

Beyer, J., R.K. Bechmann, I.C. Taban, E. Aas, W. Reichert, E. Seljeskog and S. Sanni. 2001. Biomarker measurements in long term exposures of a model fish to produced water components (PAHs and alkylphenols) Report AM-01/007. 27 pp.

Bhattacharya, A. and S. Bhattacharya. 2007. Induction of oxidative stress by arsenic in *Clarias batrachus*: Involvement of peroxisomes. Ecotoxicol. Environ. Safety. 66: 178–187.

Blechinger, S.R., J.T. Warren, Jr. J.Y. Kuwada and P.H. Krone. 2002. Developmental toxicology of cadmium in living embryos of a stable transgenic zebrafish line. Environ. Health Perspect. 110(10): 1041–1046.

Bohne-Kjersem, A., N. Bache, S. Meier, G. Nyhammer, P. Roepstorff, Ø. Sæle, A. Goksøyr and B.E. Grøsvik. 2010. Biomarker candidate discovery in Atlantic cod (*Gadusmorhua*) continuously exposed to North Sea produced water from egg to fry. Aquatic Toxicol. 96 (4): 280–289.

Cao, L., W. Huang, J. Liu, X. Yin and Sh. Dou. 2010. Accumulation and oxidative stress biomarkers in Japanese flounder larvae and juveniles under chronic cadmium exposure. Comp. Biochem. Physiol. 151C(3): 386–392.

Cao, L.,W. Huang, X. Shan, Z. Ye and Sh Dou. 2011. Tissue-specific accumulation of cadmium and its effects on antioxidative responses in Japanese flounder juveniles. Environ. Toxicol. Pharmacol. 33(1): 16–25.

Chan, W.K. and K.M. Chan. 2012 Disruption of the hypothalamic-pituitary-thyroid axis in zebrafish embryo–larvae following waterborne exposure to BDE-47, TBBPA and BPA. Aquatic Toxicol. 108: 106–111.

Chen, O., L. Yu, L. Yang and B. Zhou. 2012. Bioconcentration and metabolism of decabromodiphenyl ether (BDE–209) result in thyroid endocrine disruption in zebrafish larvae Aquatic. Toxicol. 110–111:141–148.

Chesalina, T.L. and I.I. Rudneva. 1998. Effect of heavy oil fractions on embryos and larvae of round goby *Neogobius mekanostomus*. Ichthyology J. 38(3): 426–429 (*in Russian*).

Chesalina, T.L., I.I. Rudneva and N.S. Kuzminova. 2000. Toxic effect of solar on fry of Black Sea *Liza saliens*. Ichthyology J. 40(3): 429–432 (*in Russian*).

Du, M., D. Zhang, Ch. Yan and X. Zhang. 2012. Developmental toxicity evaluation of three hexabromocyclododecane diastereo isomers on zebrafish embryos. Aquatic Toxicol. 112–113: 1–10.

Ercal, N., H. Gurer-Orhan and N. Aykin-Burns. 2001. Toxic metals and oxidative stress Part I: mechanisms involved in metal induces oxidative damage. Current Topics in Medicinal Chemistry. 1: 529–539.

Fallahtafti, Sh., T. Rantanen, R.S. Brown, V. Snieckus and P.V. Hodson. 2012. Toxicity of hydroxylated alkyl-phenanthrenes to the early life stages of Japanese medaka (*Oryzias latipes*) Aquatic Toxicol. 106–107: 56–64.

Fang, X., W. Dong, K. Thornton and C.L. Willett. 2010. Benzo[a]pyrene effects on glycine N-methyltransferase mRNA expression and enzyme activity in *Fundulus heteroclitus* embryos. Aquatic Toxicol. 98(2): 130–138.

Fang, Ch., X. Wu, O. Huang, Y. Liao, L. Liu, L. Qiu, H. Shen and S. Dong. 2012. PFOS elicits PFOS elicits transcriptional responses of the ER, AHR and PPAR pathways in *Oryzias melastigma* in a stage-specific manner. Aquatic Toxicol. 106–107: 9–19.

Farina, O., R. Ramos, C. Bastidas and E. Carcia. 2008. Biochemical responses of cnidarian larvae to mercury and benzo(a)pyrene exposure. Bull. Environ. Contamin. Toxicol. 81: 553–557.

Finn, R.N. 2007. The physiology and toxicology of salmonid eggs and larvae in relation to water quality criteria. Aquatic Toxicol. 81: 337–354.

Fleming, C.R. and R.T. Di Giulio. 2011. The role of CYP1A inhibition in the embryotoxic interactions between hypoxia and polycyclic aromatic hydrocarbons (PAHs) and PAH mixtures in zebrafish (*Danio rerio*). Ecotoxicology. 20: 1300–1314.

Fortin, M.-G., C.M. Coullard, J. Pellerin and M. Lebeuf. 2008. Effects of salinity on sublethal toxicity of atrazine to mummichog (*Fundulus heteroclitus*) larvae. Mar. Environ. Res. 65(2): 158–170.

Goksoyr, A., J. Beyer, E. Egaas, B.E. Grosvik, K. Hylland, M. Sandvik and J.U. Skaare. 1996. biomarker responses in flounder (*Platichthys flesus*) and their use in pollution monitoring. Mar. Pollut. Bull. 33: 36–45.

Gonul, L.T. and F. Kucuksezgin. 2007. Mercury accumulation and speciation in the muscle of red mullet (*Mullus barbatus*) and annular sea bream (*Diplodus annularis*) from Izmir Bay (Eastern Aegen). Mar. Pollut. Bull. 54: 1962–1967.

Hamilton, S.J. 2004. Review of selenium toxicity in the aquatic food chain. Sci. Total Environ. 326: 1–31.

Hamilton, S.J., K.M. Holleyb, K.J. Buhla and F.A. Bullarda. 2005. Selenium impacts on razorback sucker, Colorado River, Colorado. II. Eggs. Ecotoxicol. Environ. Safety. 61: 32–43.

Hampel, M., J.B. Ortiz-Delgado, I. Moreno-Garrido, C. Sarasquete and J. Blasco. 2004. Sublethal effects of linear alkylbenzenesulphonate on larvae of the seabream (*Sparus aurata*): histological approach. Histol. Histopathol. 19: 1061–1073.

Hatlen, K., C.A. Sloan, D.G. Burrows, T.K. Collier, N.L. Scholz and J.P. Incardona. 2010. Natural sunlight and residual fuel oils are an acutely lethal combination for fish embryos. Aquatic Toxicol. 99(1): 56–64.

He, Ch., Z. Zuo, X. Shi, L. Sun and Ch. Wang. 2012. Pyrene exposure influences the thyroid development of *Sebastiscus marmoratus* embryos. Aquatic Toxicol. 124–125: 28–33.

Hopkins, W.A., C.P. Tatara, H.A. Brant and C.H. Jagoe. 2003. Relationships between mercury body concentrations, standard metabolic rate and body mass in castem mosquitofish *Gambusia holbrooki* from three experimental pollution. Environ. Toxicol. Chem. 22: 586–590.

Huang, W., L. Cao, J. Liu, L. Lin and Sh. Dou. 2010a. Short-term mercury exposure affecting the development and antioxidant biomarkers of Japanese flounder embryos and larvae. Ecotoxicol. Environ. Safety. 73(8): 1875–1883.

Huang, W., L. Cao, Z. Ye, X. Xuebo Yin and Sh. Dou. 2010b. Antioxidative responses and bioaccumulation in Japanese flounder larvae and juveniles under chronic mercury exposure. Comp. Biochem. Physiol. 152C(1): 99–106.

Hutchinson, Th.H., G.T. Ankley, H. Segner and Ch.R. Tyler. 2006. Screening and Testing for Endocrine Disruption in Fish—Biomarkers As "Signposts," Not "Traffic Lights," in Risk Assessment. Environ Health Perspect. 114(S–1): 106–114.

Jeffree, R.A., F. Oberhansil and J.I. Teyssie. 2008. The accumulation of lead and mercury from seawater and their depuration by eggs of the spotted dogfish *Scyliorhinus canicula* (Chondrichthys). Arch. Environ. Contamin. Toxicol. 55: 451–461.

Karakoc, F.T., A. Tuli and A. Hewer. 1998. Adduct distribution in Piscine DNA: south-eastern Black Sea. Mar. Pollut. Bull. 36(9): 696–704.

Klumpp, D.W., C. Humphry, H. Huasheng and F. Tao. 2002. Toxic contaminants and their biological effects in coastal waters of Xiamen, China.: II Biomarkers and embryo malformation rates as indicators of pollution stress in fish. Mar. Pollut. Bull. 44(8): 761–769.

Knight, D.P., D. Feng and M. Stewart. 1996. Structure and function of selachian egg case. Biol. Rev. 71: 81–111.

Kong, X., Sh. Wang, H. Jiang, G. Nie and X Li. 2012. Responses of acid/alkaline phosphatase, lysozyme, and catalase activities and lipid peroxidation to mercury exposure during the embryonic development of goldfish *Carassius auratus* Aquatic Toxicol. 120–121; 119–125.

Kubrak, O.I., V.V. Husak, B.M. Rovenko, H. Poigner, M.A. Mazepa, M. Kriews, D. Abele and V.I. Lushchak. 2012. Tissue specificity in nickel uptake and induction of oxidative stress in kidney and spleen of goldfish *Carassius auratus*, exposed to waterborne nickel. Aquatic Toxicol. 118–119: 88–96.

Kuzminova, N.S. 2003. The influence of different concentration of sewage on marine organisms. Abstracts Int. Conf. "Space and Biosphere" Partenit, Ukraine, Sept. 28 –Oct. 4. Simpheropol. 164 (*in Russian*).

Kuzminova, N.S., V.G. Shaida and I.I. Rudneva. 2004. The influence of municipal wastewater on Black Sea fish larvae. Russian Physiol. J. 90(8): 285 (*in Russian*).

Lavado, R., D. Shi and D. Schlenk. 2012. Effects of salinity on the toxicity and biotransformation of ʟ-selenomethionine in Japanese medaka (*Oryzias latipes*) embryos: mechanisms of oxidative stress. Aquatic Toxicol. 108: 18–22.

Lesser, M.P. 2006. Oxidative stress in marine environments: biochemistry and physiological ecology. Ann. Review Physiol. 68: 253–278.

Li, H.C., Q. Zhou, Y. Wu, J. Fu, T. Wang and G. Jiang. 2009. Effects of waterborne nano-iron on medaka (*Oryzias latipes*): Antioxidant enzymatic activity, lipid peroxidation and histopathology. Environ. Safety. 72: 3684–3692.

Livingstone, D.R. 2001. Contaminant-stimulated reactive oxygen species production and oxidative damage in aquatic organisms. Mar. Pollut. Bull. 42: 656–665.

Livingston, D.R., P. Lemayre and A. Matthews. 1995. Assessment of the impact of organic pollutants on goby (*Zosterisessor ophiocephalus*) and mussel (*Mytilus galloprovincialis*) from Venice lagoon, Italy. Biochemical studies. Mar. Environ. Res. 39(1-4): 235–240.

Loro, V.L., M.B. Jorge, K.R. da Silva and Ch. M. Wood. 2012. Oxidative stress parameters and antioxidant response to sublethal waterborne zinc in a euryhaline teleost *Fundulus heteroclitus:* Protective effects of salinity. Aquatic Toxicol. 110–111: 187–193.

Lushchak, V.I. 2011. Environmentally induced oxidative stress in aquatic animals. Aquatic Toxicol. 1: 13–30.

Maceda-Veiga, A., M. Monroy, G. Viscor and A. Sostoa. 2010. Changes in non-specific biomarkers in the Mediterranean barbel (*Barbu smeridionalis*) exposed to sewage effluents in a Mediterranean stream (Catalonia, NE Spain). Aquatic Toxicol. 100(3): 229–237.

Maenpaa, K.A., O.-P. Penttinen and J.V.K. Kukkonen. 2004. Pentachlorophenol (PCP) bioaccumulation and effect on heat production on salmon eggs at different stages of development. Aquatic Toxicol. 68(1): 75–85.

Marcovecchio, J.E., V.J. Moreno and A. Perez. 1991. Metal accumulation in tissues of shark from the Bahia Blanca estuary, Argentina. Marine Environ. Res. 31: 263–274.

Mathews, T. and N.S. Fisher. 2009. Dominance of dietary intake of metal in marine elasmobranchs and teleost fish. Sci. Total Environ. 407: 5156–5161.

McCormick, J.M., M.S. Paiva, M.M. Häggblom, K.R. Cooper and L.A. White. 2010. Embryonic exposure to tetrabromobisphenol A and its metabolites, bisphenol A and tetrabromobisphenol A dimethyl ether disrupts normal zebrafish (*Danio rerio*) development and matrix metalloproteinase expression. Aquatic Toxicol. 100 (3): 255–262.

McElroy, A., C. Clark, T. Duffy, B. Cheng, J. Gondek, M. Fast, K. Cooper and L. White. 2012. Interactions between hypoxia and sewage-derived contaminants on gene expression in fish embryos. Aquatic Toxicol. 108: 60–69.

Mieiro, C.L., M.E. Pereira, A.C. Duarte and M. Pacheco. 2011. Brain as a critical target of mercury in environmentally exposed fish (*Dicentrarchus labrax*)—bioaccumulation and oxidative stress profiles.) Aquatic Toxicol. 103(3-4): 233–240.

Misra, V., V. Kumar, S.D. Pandey and P.N. Viswanathan. 1991. Biochemical alterations in fish fingerlings (*Cyprinus carpio*) exposed to sublethal concentrations of linear alkyl benzene sulphonate. Arch. Environ. Cont. Toxicol. 21: 514–517.

Neff, J.M. 1997. Ecotoxicology of arsenic in the marine environment. Environ. Toxicol. Chem. 16: 917–927.

Norman, A., H. Borjeson, F. David, B. Tienpont and L. Norrgren. 2007. Studies of uptake, elimination, and late effects in Atlantic Salmon (*Salmo salar*) dietary exposed to di-2-ethylhexyl phthalate (DEHP) during early life. Arch. Environ. Contam. Toxicol. 52: 235–242.

Oliveira, R., I. Domingues, C.K. Grisolia and A.M. Soares. 2009. Effects of triclosan on zebrafish early-life stages and adults. Environ. Sci. Pollut. Res. 16: 679–688.

Orrego, R., J. Guchardi, L. Beyger, R. Krause and D. Holdway. 2011. Comparative embryotoxicity of pulp mill extracts in rainbow trout (*Oncorhynchus mykiss*), American flagfish (*Jordanella floridae*) and Japanese medaka (*Oryzias latipes*) Aquatic Toxicol. 104(3-4): 299–307.

Osterauer, R., H.-R. Köhler and R. Triebskorn. 2010. Histopathological alterations and induction of hsp70 in ramshorn snail (*Marisa cornuarietis*) and zebrafish (*Danio rerio*) embryos after exposure to PtCl$_2$. Aquatic Toxicol. 99(1): 100–107.

Otte, J.C., A.D. Schmidt, H. Hollert and Th. Braunbeck. 2010. Spatio-temporal development of CYP1 activity in early life-stages of zebrafish (*Danio rerio*). Aquatic Toxicol. 100(1): 38–50.

Pait, A.S. and J.O. Nelson. 2003. Vitellogenesis in male *Fundulus heteroclitus* (killifish) induced by selected estrogenic compounds. Aquatic Toxicol. 64: 331–342.

Pane, E.F., C. Smith, J.C. McGreer and C.M. Wood. 2003. Mechanism of acute and chronic waterborne nickel toxicity in the freshwater cladoceran *Daphnia magna*. Environ. Sci. Technol. 37: 4382–4389.

Pauka, L.M., M. Maceno, S.C. Rossi and H.C. Silva de Assis. 2011. Embryotoxicity and biotransformation responses in zebrafish exposed to water-soluble fraction of crude oil. Bull. Environ. Contam. Toxicol. 86: 389–393.

Penttinen, O.P., J.O. Honkanen, K. Sorsa and J.V.K. Kukkonen. 2005. Can aquatic pollutants cause specific endocrinological and metabolic responses in salmon (*Salmo salar* m. Sebago) embryos? A direct calorimetry study. Verhandlungen International Vereinigung Limnology. 29: 945–948.

Penttinen, O.-P. and J.V.K. Kukkonen. 2006. Body residues as dose for sublethal responses in alevins of landlocked salmon (*Salmo salar* m. Sebago): a direct calorimetry study. Environ. Toxicol. Chem. 25(4): 1088–1093.

Peters, L.D., C. Porte C., J. Albaiges and D.R. Livingstone. 1994. 7-Ethoxyresorufin O-deethylase (EROD) and antioxidant enzyme activities in larvae of sardine (*Sardina pilchardus*) from the North Coast of Spain. Mar. Pollut. Bull. 28(5): 299–304.

Peters, L.D., C. Porte and D.R. Livingstone. 2001. Variation of antioxidant enzyme activities of Sprat (*Sprattus sprattus*) larvae and organic contaminant levels in mixed zooplankton from the Southern North Sea. Mar. Pollut. Bull. 42(11): 1087–1095.

Porte, S., E. Escartin, L.M. Garcia, M. Sole and J. Albaiges. 2000. Xenobiotic metabolizing enzymes and antioxidant defences in deep-sea fish: relationship with contaminant body burden. Marine Ecological Progressive Series. 192: 259–266.

Richards, J.G. 2010. Metabolic rate suppression as a mechanism for surviving environmental challenge in fish. Progress in Molecular and Subcellular Biology. 49: 113–135.

Rodrigues, A.P.C., P.O. Macie, L.C.C. Pereira da Silva, C. Albuquerque, A.F. Inácio, M. Freire, A.R. Linde, N.R.P. Almosny, J.V. Andreata, E.D. Bidone and Z.C. Castilhos. 2010. Biomarkers for Mercury Exposure in Tropical Estuarine Fish. J. Braz. Soc. Ecotoxicol. 5 (1): 9–18.

Rudneva, I.I. 1993. Mercury effect on biochemical parameters of fish. In: Ichthyophauna of Black Sea bays under anthropogenic impact. Kiev: Naukova Dumka Publ. 71–77 (*in Russian*).

Rudneva, I.I. 1998. The biochemical effects of toxicants in developing eggs and larvae of Black Sea fish species. Marine Environ. Res. 46(1): 499–500.

Rudneva, I.I. 2011. Ecotoxicological Studies of the Black Sea Ecosystem. The Case of Sevastopol Region. NY: Nova Science Publishers, Inc. 62 pp.

Rudneva, I.I. and N.V. Jerko. 1993. The effects of PCB on protein and lipid metabolism in blood serum of Black Sea scorpion fish. Reports of Ukrainian Academy of Sciences. 11. 157–161 (*in Russian*).

Rudneva, I.I., T.L. Chesalina, V.G. Shaida and N.F. Shevchenko. 1998. Morphology and heat production in atherina larvae (*Atherina hepsetus* L.) from contaminated and non-contaminated regions. Marine Ecology. 47: 33–36 (*in Russian*).

Rudneva, I.I., T.L. Chesalina and N.S. Kuzminova. 2000. The response of o Black Sea *Liza saliens* fry on mazut pollution. Ecology. 4: 304–306 (*in Russian*).

Rudneva, I.I. and E. Petzold-Bradley. 2001. Environmental and security challenges in the Black Sea region. In: Environment Conflicts: Implications for Theory and Practice, (E. Petzold-Bradley, A. Carius and A. Vince eds.). pp. 189–202. Netherlands: Kluwer Academic Publishers.

Rudneva, I.I., I.N. Zalevskaya and V.G. Shaida. 2003. The effects of polychlorinated biphenyls on *Atherina hepsetus* larvae. Ichthiology J. 43(2): 272–276 (*in Russian*).

Rudneva, I.I. and I.N. Zalevskaya. 2004. Atherina larvae (*Atherina hepsetus*) as bioindicators of pollution in coastal waters of Black Sea. Ecology. 2: 107–112 (*in Russian*).

Rudneva, I.I., I.N. Zalevskaya, N.S. Kuzminova and E.G. Savkina. 2004a. Effect of fungicide cuprocsat on Black Sea *Atherina* larvae. Agroiecological J. 3: 83–86 (*in Russian*).

Rudneva, I.I., V.G. Shaida, N.S. Kuzminova and I.E. Kutsuruba. 2004b. Analysis of the cyfose toxicity with the use of *Artemia salina* nauplia. Agroecological J. 3: 57–62 (*in Russian*).

Rudneva, I.I., V.G. Shaida, N.S. Kuzminova and I.E. Kuzuruba. 2005. *Artemia salina* L. as test-organism in the evaluation of the toxicity of fungicide cuprocsat. Biology of the Indoor Waters. 3: 104–109.

Rudneva, I.I. and V.G. Shaida. 2006. Metabolic strategy in the early life of some Black Sea fish species measured by microcalorimetry method. XIVth Conference The Amber ISBC. Abstr. Sopot, Poland, June 2–6, 2006. 74.

Rudneva, I.I. and V.G. Shaida. 2011. Response of fish early life stages on oil pollution of marine environment. In: Modern problems of aquatic toxicology. Petrozavodsk, Petrozavodsk State University Publ. 124–127.

Rudneva, I.I. and V.G. Shaida. 2012. Metabolic rate of marine fish in early life and its relationship to their ecological status. In: Fish Ecology. (Ed. P. Gempsy) Nova Science Publishers. New York, pp. 1–29.

Sánchez-Marín, P., J. Santos-Echeandía, M. Nieto-Cid, X. Álvarez-Salgado and R. Beiras. 2010. Effect of dissolved organic matter (DOM) of contrasting origins on Cu and Pb speciation and toxicity to *Paracentrotus lividus* larvae. Aquatic Toxicol. 96(2): 90–102.

Sevcikova, M., H. Modra, A. Slaninova and Z. Svobodova. 2011. Metals as a cause of oxidative stress in fish: a review. Veterinarni Medicina. 56(11): 537–546.

Sole, M., S. Rodriguez, V. Papiol, F. Maynou and J.E. Cartes. 2009. Xenobiotic metabolism markers in marine fish with different trophic strategies and their relationship to ecological variables. Comp. Biochem. Physiol. 149C: 83–89.

Storelli, M.M., R. Giacominelli-Stuffler, A. Storelli and G.O. Macotrigiano. 2005. Accumulation of mercury, cadmium, lead and arsenic in swordfish and blufin tuna from the Mediterranean Sea: A comparative study. Mar. Pollut. Bull. 50: 903–1007.

Tatara, C.P., M. Mulvey and M.C. Newman. 1999. Genetic and demographic responses of mosquitofish (Gambusia holbrooki) exposed to mercury from multiple generations. Environ. Toxicol. Chem. 18: 2840–2845.

Turoczy, N.J., L.B. Laurenson, G. Allinson, M. Nishikawa, D.F. Lambert, C. Smith, J.P.E. Cottier, B. Irvine and F. Stagnitti. 2000. Observations on metal concentrations in three species of shark (*Deania calcea, Centroscymnus crepidater* and *Centroscymnus owstoni*) from Southeastern Australian Waters. J. Agricultural Food Chemistry. 48: 4357–4364.

Utvik, T.I.R. 1999. Chemical characterisation of produced water from four offshore oil production platforms in the North Sea. Chemosphere. 39(15): 2593–2606.

Valko, M., H. Morris and M.T.D. Cronin. 2005. Metals, toxicity and oxidative stress. Current Medicinal Chemistry. 12: 1161–1208.

Vieira, M.C., R. Torronteras, F. Córdoba and A. Canalejo. 2011. Acute toxicity of manganese in goldfish *Carassius auratus* is associated with oxidative stress and organ specific antioxidant responses. Ecotoxicol. Environ. Safety. 78: 212–217.

Wright, D.A. 1995. Trace metal and major ion interactions in aquatic animals. Mar. Pollut. Bull. 31(1-3): 8–18.

Wu, S.M., Y.-C. Ho and M.J. Shin. 2007. Effects of Ca^{2+} or Na^+ on metallothionein expression in Tilapia larvae (*Oreochromis massambicus*) exposed to cadmium or copper. Arch. Environ. Contamin. Toxicol. 52: 229–234.

Wu, S.M., C.F. Weng., J.C. Hwang, C.J. Huang and P.P. Hwang. 2000. Metallothionein induction in early larval stages of tilapia (*Oreochromis massambicus*). Physiol. Zool. 5: 531–537.

Wu, Y.D., L. Jiang, Z. Zhou, M.H. Zheng, J. Zhang and Y. Liang. 2008. CYP1A/regucalcin gene expression and edema formation in zebrafish embryos exposed to 2,3,7,8-tetrachlorodibenzo-p-dioxin. Bull. Environ. Contam. Toxicol. 80: 482–468.

Wu, Y. and O. Zhou. 2012. Dose- and time-related changes in aerobic metabolism, chorionic disruption, and oxidative stress in embryonic medaka (*Oryzias latipes*): Underlying mechanisms for silver nanoparticle developmental toxicity. Aquatic Toxicol. 124–125: 238–246.

Zalevskaya, I.N., Z.S. Matveeva and I.I. Rudneva. 2004. Evaluation of the fungicide cuprocsat toxicity on *Artemia salina*. Agroecological J. 3: 75–78 (*in Russian*).

Zlabek, V., T. Randak, J. Kolarova, Z. Svobodova and H. Kroupova. 2009. Sex differentiation and vitellogenin and 11-ketotestosterone levels in chub, *Leuciscus cephalus* L., exposed to 17 β-estradiol and testosterone during early development. Bull. Environ. Contam. Toxicol. 82: 280–284.

CHAPTER 5

Stress Biomarkers in Fish Embryos and Larvae and Climate Changes

Climate change has had significant impacts on the aquatic ecosystems, including changes of hydrological and hydrochemical properties of water and sediments and biota (Thomas et al., 2004). Of particular concern are estuarine and coastal environments, which is fed by diminishing snow pack runoff leading to gradual increases in salinity. Salinity enhances the acute toxicity of several agricultural chemicals in fish through augmented biochemical activation catalyzed by enzymes that are induced during hypersaline acclimation. In addition, increase in biogen concentration together with the increasing temperature stimulates microalgae bloom and eutrophication and anoxia/hypoxia which provoke very negative consequences for water bodies, fish and invertebrates. Given the rapid changes taking place in the world's waterways, environmental modification of toxicological pathways should be a significant focus of the research community as the toxicity of multiple xenobiotics may be enhanced (Schlenk and Lavado, 2011).

5.1 Eutrophication Effects

Eutrophication is a complex process, which occurs both in fresh and marine waters, where excessive development of certain types of algae disturbs the aquatic ecosystems and becomes a threat for animal and human health. The primary cause of eutrtophication is an excessive concentration of nutrients originating from agricultural and sewage treatment. Eutrophication is commonly linked with algae blooms, "red tides", "green tides", fish kills, inedible shellfish, blue algae and public health threats. The main cause of eutrophication is the large input of nutrients into water ecosystem and the main effect is the imbalance in the food web that results in high levels of phytoplankton biomass in stratified water bodies. This can lead to algal blooms. The direct consequence is an excess of oxygen consumption near the bottom of the water body. The enrichment of water by nutrients can be of natural origin and anthropogenic input such as domestic, industrial and agricultural sewage and, runoff and erosion (Borisova et al., 2005). The main consequences of eutrophicaion are increased abundance of plankton algae, reduced Secchi depth, lack of oxygen, development of hypoxic and anoxic zones, changes phyto- and zooplankton composition, reduction in biodiversity and fish resources and in some cases ecosystem degradation (Rudneva and Petzold-Bradley, 2001).

Eutrophication is in part a natural process whereby nitrogen and phosphorus accumulate in water and water enters into the aquatic ecosystems from surrounding land. However, various forms of human activity cause this accumulation to increase, leading to a eutrophication process with potentially drastic effects on biota. Examples of consequences include an increase of vegetation covering shallow bays, intensive algal blooms, turbid water and effects on higher trophic levels such as fish. Intensive eutrophication leads to an increased production of algae mass which when sedimented at bottom leads to oxygen deficiency, as oxygen is used when dead plants and animals decompose. Eutrophication is a global problem that is often manifested in coastal areas where human activity leads, in various ways, to increased nitrogen and phosphorus levels. Depending upon a number of chemical, hydrological and biological factors, a sea can "react" in different ways to large accumulations of nitrogen and phosphorus and develop "hot ecological points". There are large regional differences as regards eutrophication, as well as local differences between coastal and open sea areas (Almesjo and Limen, 2009). The structure of the habitat is usually crucial for growth and survival of early life stages of fish and invertebrates. Presently, some nursery areas of fish larvae are changing due to eutrophication and in some cases the structure of the habitat was probably too open for newly hatched larvae (Engström-Öst et al., 2007).

The effects of eutrophication on fish populations are complex. Initially, eutrophication may have a positive impact on fish communities since the availability of food tends to increase. In the long-term, however, eutrophication may cause oxygen depletion, alter the characteristics of vital habitats, and cause changes in the species composition of prey, which may cause a decline in fish populations. The whole ecosystem must be considered when tackling the problems of eutrophication. Recent studies show that the absence of predatory fish such as cod can worsen the effects of eutrophication. A decline of predatory fish may (in the same way as inincreased discharge of nutrients) lead to a higher abundance of phytoplankton or macro-algae and in extension cause oxygen depletion and dead zones (Almesjo and Limen, 2009).

The anthropogenic modification of nitrogen circulation in biosphere caused by human activity leads to high distribution and accumulation of nitrogen compounds in the environment, including marine ecosystems. Direct sewage into the sea from coastal settlements containing heavy loads of fertilizer and human waste is the major contributing factor for the degradation of marine environment. Domestic and agricultural sewage caused eutrophication and formation nitrosamines (NA) in water, sediments and in aquatic biota. The formation of NA is relatively easy from nitrogen containing precursors such as nitrogen oxides, nitrates, nitrites, amines and amides. Exogenous NA synthesis was detected in water, soil, air and endogenous was shown in biota. The NA level in environment depends upon the circulation of their nitrogen precursors, their distribution in environment, microorganisms' metabolism. The anthropogenic impact on ecosystem including water and air pollution together with solar radiation disturbs the normal nitrogen balance in biosphere and induces the formation of NA compounds (Rudneva et al., 2008).

In recent years NA have been reported as the most potent groups of known cancerogenes. They cause cancer, mutations, embryonic defects and anomalies of development in 40 species of animals including aquatic organisms (Rubenchik, 1990; Winston et al., 1992). The precursors of NA compounds (nitrogen oxides, nitrates, nitrites, amines and amides) are highly distributed in the environment. Besides that NA is formed from some pesticides and drugs. Interaction between nitrite and certain amine compounds in marine water might be expected to result in the formation of NA. Bacteria and microorganisms also play an important role in this process. They might reduce nitrate to nitrite (the main precursor of NA) in some cases. In water that receives excess nitrogenous waste, an imbalance between bacteria nitrification and denitrification can occur, eventually leading to nitrite accumulation. It acts to increase water nitrite that can accumulate to high levels that are toxic to aquatic animals. In this case NA pose potential hazards to the health of fish and people because they are transferred via food chains (Deane and Woo, 2007). Fertilized and hatching eggs are the most sensitive to toxic effects of exogenous NA. In laboratory conditions it was demonstrated that NA in high concentrations decreased the hatching of fertilized eggs and therefore through their effect on eggs hatchability may reduce fish populations along with increasing aquatic eutrophication (Bieniarz et al., 1996).

Hypoxia and anoxia effects

When organic matter that has sedimented at the bottom decomposes, oxygen is consumed, which can lead to a lack of oxygen in the sediment and bottom water. The lack of oxygen becomes especially extensive when the abundance of organic matter is high and the water column mixing is poor. The effects of hypoxia on aquatic animals living in an ecosystem experiencing low oxygen levels (< 2 mg l^{-1}) coincide with a reduction of demersal fish and death of benthic fauna (Hassell et al., 2009). Fish early stages are very sensitive to eutrophication because hypoxia/anoxia, microalgae toxins and high levels of nutrients cause negative biological events in their habitats and, specially, in coastal and estuarine areas which are the spawning sites of fish and invertebrates. A lack of oxygen can cause hydrogen sulphide to form, which is highly toxic for most organisms. A complete lack of oxygen also disturbs the nitrogen cycle. The potential for denitrification, when nitrate is converted into nitrogen gas and transported out of the system, is weakened. Anoxic sediments also release phosphorus, which in conditions of oxygen-saturation, stays trapped in the sediment (Conley et al. 2002).

Lack of oxygen is considered as one of the several main threats to coastal ecosystems globally and in coastal zones, the number of dead sites has increased dramatically since 1960, and today more than 400 areas are affected by a serious lack of oxygen. The detrimental effects of lack of oxygen on marine environments is widespread today and the exponential increase in the number of anoxic environments in a short period is alarming (Diaz and Rosenberg, 2008). Hypoxia is a common feature of eutrophic water bodies. Hypoxia occurs in bottom layers because photosynthesis requires light which is lacking in bottom unmixed waters (Randall et al., 2004).

The major consequence of eutrophication concerns the availability of oxygen. The imbalance of the processes of oxygen production and utilization in eutrophic water bodies leads to loss of oxygen in water and all biota disappear. The major effect of hypoxia on an individual is to reduce exercise capacity. Fish respond to hypoxia by inhibiting feeding and reproduction and moving to lower temperatures, all of which lowers energy expenditure. Oxygen delivery is augmented by increasing gill ventilation and hemoglobin content, and oxygen affinity. Aerobic metabolism is down-regulated whereas anaerobic metabolism is up-regulated. Steroid metabolism is reduced. Liver cells go into cell cycle arrest. Cell growth and aerobic metabolism genes are down-regulated but genes associated with anaerobic metabolism are up-regulated. Uncoupling proteins are up-regulated and may play a role in reducing mitochondrial activity during hypoxia (Randall et all., 2004).

The presence of gross abnormalities in eelpout broods has been suggested to be a useful biomarker of the impact of hazardous substances on fish reproduction in the marine environment as chronic exposure to various substances has the potential to induce severe developmental defects in fish embryos and larvae. Additionally, severe oxygen depletion suggests that examination of broods in the eelpout may include not only impact of hazardous substances but also effect of eutrophication-related problems on fish reproduction in the marine environment (Strand et al., 2004). Prolonged exposure to hypoxia results in behavioral or physiological changes such as increased ventilation frequency and cardiac output in both fish and crustaceans. While these responses are valuable indices of low oxygen conditions, they cannot serve as effective hypoxia biomarkers since they involve *in situ* measurements of the organisms and are not practical for monitoring. Genes that respond to hypoxia can potentially serve as indicators of hypoxic stress (Strand et al., 2004).

Hypoxia is an endocrine disruptor, and can lower the levels of sex hormones testosterone and estradiol in adult carps (by 77–91%). Disturbances of sex hormones retarded gonad development in both male and female fish, leading to a decline in sperm motility, decrease in fertilization success (from 99.4% to 55.5%), hatching rate (from 98.8% to 17.2%) and larval survival (from 93.7% to 46.4%). Overall, the survival of eggs to larvae decreased from 92.3% in the normoxic group to only 4.4% in the hypoxic group. Hypoxia is a teratogen, and can significantly increase % malformation (by 77.4%) during fish embryonic development. Disruption of apoptotic pattern was clearly evident at 24 hpf, which may be a major cause of malformation. Hypoxia can also delay embryonic development, and upset the balance of testosterone and estradiol at very early developmental stages, implicating that subsequent sexual development may also be affected. Taken together, our results imply that hypoxia may reduce species fitness and threaten the sustainability of natural fish populations over large areas on a global scale. Compared with the normoxic control, number of apoptotic cells in the tail of hypoxic embryos were significantly reduced (−63.7%, p < 0.01). In contrast, a significantly higher percentage (+116%, p < 0.05) of apoptotic cells was found in the brain of hypoxic embryos as compared to control embryos. The authors suggested that the apoptotic pattern in zebrafish embryos was altered by hypoxia (Wu et al., 2004).

Reproduction and endocrine function are particularly susceptible to interference by environmental hypoxia exposure. Marked impairment of reproductive function and endocrine disruption was observed in individuals of hypoxic-tolerant estuarine teleost, Atlantic croaker collected from hypoxic sites in East Bay, Florida and Mobile Bay, Alabama. The production of mature oocytes and sperm (gametogenesis), as well as sex steroid and vitellogenin levels in the blood, were significantly lower in croaker from the hypoxic sites in East Bay compared to the values in fish collected from the adjoining normoxic Pensacola Bay, whereas gonadal HIF-1alpha and HIF-2alpha mRNA expressions were significantly elevated in fish from the hypoxic sites. Reproductive and endocrine functions were also impaired in female croaker collected from the hypoxic zone off the Louisiana coast. The production of mature oocytes (fecundity) was significantly decreased in fish collected from the hypoxic site compared to that observed at the normoxic site and this was associated with declines in circulating sex steroid and vitellogenin levels and gonadotropin releasing hormone mRNA expression in the hypothalamus. Tissue expressions of HIF-2 αmRNA and protein were significantly increased in croaker collected at the hypoxic site. The authors concluded that assessment of HIFα(s) expression and reproductive/ endocrine functions are promising biomarker of exposure to hypoxia and its potential long-term impacts on fish populations, respectively (Thomas and Rahman, 2009).

Fish embryos were used to evaluate the interaction among common environmental and chemical stressors found in urban coastal environments, namely hypoxia, aryl hydrocarbon receptor (AhR) agonists, and estrogenic compounds. Biomarkers of exposure to these stressors, cytochrome P4501A (CYP1A), estrogen receptor alpha (ERα), brain cytochrome P450 aromatase (CYP19A2 or AromB), and hypoxia inducible factor 1α (HIF-1α) mRNA expression were examined using qRT-PCR simultaneously in embryos of two species, the Atlantic killifish, *Fundulus heteroclitus,* and the zebrafish *Danio rerio.* Embryos of both species were exposed to the model CYP1A inducer β-naphthoflavone (BNF) or 17-β estradiol (E2) under either normoxic or hypoxic (5% oxygen atmosphere) conditions and harvested prior to hatch at 9 days post fertilization (dpf) for the killifish, and 48h post fertilization (hpf) for the zebrafish. BNF significantly induced CYP1A expression in embryos of both species with killifish embryos being more responsive (700-fold > control) than zebrafish embryos (7-100-fold > control). AromB was also significantly influenced by treatment, but to a lesser extent, with mean expression levels increased by less than two-fold over control values in response to E2, and in one case up-regulated by BNF. ERα and HIF-1α were constitutively expressed in embryos of both species, but expression was unaffected by exposure to either BNF or E2. Hypoxic conditions down-regulated AromB expression strongly in killifish but not in zebrafish embryos. An interactive effect between hypoxia and BNF on several of the genes evaluated was observed. The results support the use of Cyp1a expression as a biomarker of AhR agonists in fish embryos, and indicate that AromB may be more responsive than ERα to estrogenic chemicals at this stage of development. Killifish embryos are generally more sensitive than zebrafish embryos at this stage of development supporting their use in environmental assessments (McElroy et al., 2012).

Both hypoxia and hyperoxia, albeit in different magnitude, are known stressors in the aquatic environment. Mirror carps (*Cyprinus carpio* L.) were exposed chronically (i.e., 30 days) to hypoxic (1.8 ± 1.1 mg O_2 l^{-1}) and hyperoxic (12.3 ± 0.5 mg O_2 l^{-1}) conditions and resultant biological responses (biomarkers) were compared between these two treatments as well as with fish held under normoxic conditions (7.1 ± 1.04 mg O_2 l^{-1}). The biomarkers determined included the activities of glutathione peroxidase

(GPx), measurement of oxidative DNA damage, hematological parameters, histopathological and ultrastructural examination of liver and gills. Specific growth rate (SGR) of the fish was also determined over the exposure period. The study suggested that while the levels of hepatic GPx were unaffected, there was a significant difference in activity in the blood plasma under different exposure conditions; the hyperoxic group showed increased GPx activity by approximately 37% compared to normoxic group and the hypoxic group showed a decrease by approximately 38% than the normoxic group. Oxidative DNA damage was significantly higher in both hypoxic and hyperoxic by approximately 25% compared to normoxic conditions. The hematological parameters showed enhanced values under hypoxic conditions. SGR of fish was significantly lowered in hypoxic by approx. 30% compared to normoxic condition and this was found to be correlated with DNA damage (R = –0.82; P = 0.02). The results obtained indicate that prolonged exposure to both hypoxic and hyperoxic conditions induce oxidative stress responses at both DNA and tissue levels, and hypoxia can result in compensatory changes in hematological and growth parameters (Mustafa et al., 2011).

Hypoxia is known to disrupt the endocrine systems in fish, retarding gonadal development and reducing success in spawning, fertilization and larval growth. In normoxia, fertilized zebrafish eggs hatch between 48 and 60 hours post fertilization and 93.8% of the eggs hatched. In hypoxia, fertilized eggs did not hatch until 96 to 260 hours post fertilization, only 4.9% hatched and the rest died. Fertilized eggs developing under hypoxia were pale, indicating a lack of pigment. Growth was retarded and there were many abnormalities. Hypoxia exposure had little effect on the expression of the aryl hydrocarbon receptor nuclear translocator or the activity of the P450scc (CYP11A) (Randall et al., 2004).

Algae bloom and their toxins

At the base of marine food chain, phytoplankton is an important indicator of changes in the sea. Additionally, some microalgae are the food of zooplankton organisms and they, alongwith fish larvae, play an important role in food chain in aquatic ecosystem. In eutrophic waters the phytoplankton composition is changed and cyanobacterial species dominate, some of them are toxic for fish and invertebrates. Larger accumulations of cyanobacteria affect the quality of food for organisms in the water column and at the bottom, once the bacteria has sedimented. In addition to the effects of the toxins cyanobacteria can produce, researchers have found various indirect effects such as behavioral alterations among fish and even changes in the surrounding plankton community, its composition and food quality (Rudneva et al., 2008).

Large accumulations of cyanobacteria in the summer can affect fish communities in different ways, partly through changes in the Secchi depth and through exposure to the toxins that are sometimes released in conjunction with heavy blooms. Toxins from cyanobacteria can accumulate in higher trophic levels through zooplankton (Engstrom-Ost et al., 2002). It is estimated that less than 1% of toxins from cyanobacteria are accumulated in higher trophic levels, but also very small amounts can affect organisms (Karjalainen et al. 2007). The reaction among smaller-sized fish may be more dramatic than among adult fish—embryonic development of herring has, for example, been seen to be negatively affected by accumulations of cyanobacteria (Ojaveer et al. 2003). Adult salmon trout can get serious liver damage from nodularin, which is one of the toxins produced by cyanobacteria. This damage has shown itself to be reversible once the fish are no longer exposed to the toxins (Kankaanpaa et al. 2002).

Accumulations of non-toxic blooms have also been shown to affect the behavior of fish. The search for food among juvenile pike has, for example, been seen to decrease during heavy blooms which may be because of both poorer visibility and that their gills are blocked (Engstrom-Ost et al., 2006, 2007). Changes in the spawning behavior of sticklebacks on account of increased turbidity have also been observed (Engstrom-Ost and Candolin, 2007). It must be regarded as uncertain whether large cyanobacterial blooms can have large-scale effects on fish and other organisms. Bearing in mind that cyanobacteria are rather large and are only grazed to a limited extent, it is, however, not very likely that they could cause damage on the same scale as dinoflagellates and other smaller, toxin-producing plankton algae.

The occurrence of heavy cyanobacterial blooms in eutrophic freshwater and marine ecosystems has been a worldwide problem. 50–75% of the cyanobacterial blooms were detected to be toxic. Of the 70+ isoforms of microcystins (MC) are identified, among them hepatotoxin microcystin-LR (MC-LR), is the most studied, toxic and common (Deng et al., 2010). Microcystin the predominant toxins of cyanobacteral blooms, caused the mortality, illness and stress in aquatic animals including fish. MCs are the potential inhibitors of protein phosphatases in hepatocytes of animals which associated with the imbalance of protein phosphorilation and may induce liver disease or necrosis. In eutrophic and hypereutrophic waters, cyanobacteria often dominates the phytoplankton. Cyanobacteria produce a wide range of natural toxins, including cyclic peptides (microcystins and nodularin), alkaloids, and lipopolysaharides (Gazenave et al., 2006). Over 70 structural variants of microcystins (MC) have been identified, most of them are highly toxic. On the other hand, in water ecosystems a variety of compounds are present, which may produce either increased toxicity (synergy) or attenuate the effects of cyanobacterial toxins. An essential component of aquatic environment is natural organic matter (NOM) which plays an important role as a photo sensitizer for indirect photolytic breakdown of cyanotoxins. Natural organic matter (NOM) can bind or integrate a wide range of organic pollutants and metals in their structure, and thereby decrease the bioavailability, and consequently the toxicity of these substrates or modify the toxicity of different harmful chemicals on several aquatic animals (Gazenave et al., 2006).

Oxidative stress induced by MC exposure is considered to be involved in the toxicity of MCs and lead to induction of biotransformation and antioxidant system which are good tools for the evaluation of toxic effects of cyanobacterial blooms on the fish liver. The response of fish biomarker on MC exposure depends on fish ecological status. The phytoplanktivorous fish possessed higher basal GSH concentrations and better correlations among major antioxidant enzymes in liver which might be responsible for their fish communities following occurrence of toxic cyanobacterial blooms and also for applicability of using phytoplanktivorous fish to counteract toxic cyanobacterial blooms in natural waters (Qiu et al., 2007).

Cyanobacterial blooms and their toxins contribute to mass mortalities of fish in eutrophic waterbodies, besides other contributing factors, such as high pH and low oxygen. MC altered the behavior of adult fish, impairing feeding and reproduction behavior, diurnal rhythmic to nocturnal (Baganz et al., 2004). Cyanobacterial crude extracts, containing environmental concentrations of cyanotoxins, caused malformations and mortalities in carp embryos, later hatching and decreased total length of yolk sac dropsy (Palikova et al., 2007). However, NOM itself also has effects on aquatic organisms. Such tensid-like action on membranes, alteration of development, growth and reproduction, modifies the biotransformation enzymes and antioxidant status of the biota (Wiegand et al., 2004).

Fish embryos and larvae are the most sensitive stages in the life cycle of teleost, being particularly sensitive to various stressors to which they might be exposed. Impact of pure microcystins (-RR and -LF) on zebrafish (*Danio rerio*) embryos were investigated individually and in combination with a NOM. The applied NOM was a reverse osmosis isolate from Lake Schwarzer See (i.e., Black Lake, BL-NOM). Teratogenic effects were evaluated through changes in embryonic development within 48 h of exposure. Detoxication activities were assessed by the activities of the following biomarkers: phase II biotransformation enzymes, soluble and microsomal glutathione S-transferase (m GST). After uptake of MC-LR by early life stages of zebrafish, their GST detoxification system reacted to this toxin, indicating metabolization of MC-LR to an excretable glutathione conjugate. The investigators showed that GST activity was increased during exposures to two MC congeners, pure MC-RR and MC-LF, while enzyme activity did not significantly change in embryos exposed to either BL-NOM or BL-NOM combined with MC-RR or MC-LF. Hence, BL-NOM attenuated toxic effects of MC-LF and MC-RR verified by less pronounced teratological effects within 24 h, in particular, as well as less rise in the activity of s-GST, when compared with embryos exposed to either pure toxins or in combination with organic matter (Gazenave et al., 2006).

Study of the toxic threshold of male and female fish to microcystins is based on different biomarkers. Japanese medaka (*Oryzias latipes*) were dietary fed. Microcystin-LR (0, 0.46, 0.85, 2.01 and 3.93 mg MC-LR/g) were fed with dry diet for 8 weeks at 25°C. The results revealed that dietary MC-LR inhibited growth at the end of 8 weeks. The survival of embryos and the RNA/DNA ratio of whole fish decreased significantly (P < 0.05) in fish fed with 3.93 mg MC-LR/g dry diet. Heat shock protein (Hsp60) expression was induced in the liver of female and male fish fed with diets containing 0.85 and 0.46 mg MC-LR/g diet, respectively. The activity of liver caspase 3/7 was significantly higher in female fish fed with 3.93 mg MC-LR/g diet and in males fed with 2.01 MC-LR mg/g dry diet than fish fed with the control diet. The threshold for inhibition of liver protein phosphatase expression was lower in female (2.01 mg/g diet) than that in male fish (3.93 mg/g diet). Histopathological examination showed significant single cell necrosis in female and male Medaka fed with diets containing 0.85 and 3.93 mg MC-LR/g diet, respectively. Based on different biomarkers, this study demonstrated that dietary MC-LR is toxic to Medaka and the effects are sex dependent (Deng et al., 2010).

The metabolism of toxins results in the formation of reactive oxygen species (ROS) and oxidative stress which significantly contribute to their toxicity. Oxidative stress was assessed by determining both the production of hydrogen peroxide and by analyzing the activities of the antioxidat enzymes, guaiacol peroxidase (POD), catalase (CAT), glutathione peroxidase (GPx), and glutathione reductase (GR) in zebrafish embryos. Hydrogen peroxide content of the embryos was elevated by all cyanotoxins as well as BL-NOM produced oxidative stress to the embryos, which in turn should be reduced by antioxidant enzymes such as CAT, POD and GPx. CAT was the only antioxidant enzyme of the zebrafish showing high activity and always following a pattern similar to that observed for H_2O_2 formation. Pure toxins or their combination with BL-NOM revealed opposite trends of hydrogen peroxide formation: pure MC-LF caused higher H_2O_2 than pure MC-RR, with the concomitant activation of the CAT, but the combination of BL-NOM with the toxins was able to diminish only oxidative stress of MC-LF, while this negative effect was further increased in combination with MC-RR. The mode action of NOM strictly depends upon its molecular characteristics, which is influenced by their origin, which can also influence the binding ability to certain toxins and might explain differences in enzyme activities. The authors suggested that BL-NOM also diminished oxidative effects caused by MC-LF; however, it failed to attenuate oxidative stress caused by MC-RR (Gazenave et al., 2006).

Detoxification implies the use of additional energy to support this defense system. Content of lipids was significantly reduced in exposed embryos following a trend similar to that obtained with teratological and enzymatic assays confirming the attenuating effect of BL-NOM. Carbohydrates show a similar trend but without significant differences between control and tested groups. Physiological responses to microcystins and NOM required energetic costs, which were compensated by the expense of the energy resources of the yolk, which in turn might affect the normal development of embryos (Gazenave et al., 2006).

Therefore, several factors such as hypoxia/anoxia/hyperoxia, toxic nitrogen compounds and microcystins affected early developmental stages of fish in euthroph areas. The survival and the success of their further development of fish in these habitats depend on the fish ability to defense against stress and to adapt to unfavorable conditions.

5.2 Increasing Water Temperature

The Earth's climate has warmed by approximately 0.6°C over the past century, and the net heat uptake by the world's ocean has been more than 20 times greater than that by the atmosphere (Levitus et al., 2005). Average global sea surface temperatures are expected to increase up to 3°C by 2100 (Meehl et al., 2007), which is predicted to change of marine biodiversity (Edwards and Richardson, 2004; Thomas et al., 2004), physiology, metabolism (Pörtner and Knust, 2007) and distribution of aquatic organisms (Perry et al., 2005). Projections of future changes suggest that the global average temperature will increase by 1.4° to 5.8°C by the end of the 21st century in comparison to 1990 level. This magnitude of change is likely to be unprecedented for at least the last 1000 years. The other problem of the consequences of climate change is connected with the depletion of ozone layer and the increase of UV-irradiation in the distinct regions of the planet (see 5.3).

Temperature is a key regulator of fish metabolism, growth and reproduction. The environment and water quality may be different for fish larvae in coastal areas and in open sea. In addition, environmental conditions may vary much for wild fish larvae and larvae in aquaculture. However, the information about the capacity of fish' early life stages to adapt to elevated temperatures is very limited but crucial to understand how marine fish eggs and larvae will respond to global warming. The investigators propose, that the stressful abiotic conditions inside eggs are expected to be aggravated under the projected near-future ocean warming, with deleterious effects on embryo survival, growth and development.

Greater feeding challenges and the lower thermal tolerance limits of the hatchlings are strictly connected to high metabolic demands associated with the planktonic life strategy and in the future, the early stages might support higher energy demands by adjusting some cellular functional properties to increase their thermal tolerance windows (Rosa and Seibel, 2008; Rosa et al., 2012). Thus, more knowledge is needed of the temperature effects on larval fish physiology and biochemistry through early life hystory. Similarly, the effects on larval fish of variations in salinity, oxygen content and pollution need further exploration.

The impact of increase in temperature on marine organisms was documented in different biological levels:

- in molecular it associated with the heat shock proteins induction;
- in physiological level the metabolism modification was observed;
- in organism the changes of behavior, season fluctuations, feeding, reproduction and development were documented;
- in population level the changes of spawning and migration were shown;
- in communities level the alterations in the links between their components were detected;
- in ecosystems and biosphere the global changes in environments and biota were documented.

Because many fish spawn in coastal waters, ocean warming will negatively impact their reproduction, hatching and larvae survival. The metabolism of marine organisms is constrained by oxygen supply at high (and low) temperatures with a progressive transition to an anaerobic mode of energy production (the "oxygen limitation of thermal tolerance" concept (Pörtner and Knust, 2007; Pörtner et al., 2004; Rosa and Seibel, 2008). The reduction in aerobic scope is not caused by lower levels of ambient oxygen but through limited capacity of oxygen supply mechanism (ventilatory and circulatory systems) to meet an animal's temperature-dependent oxygen demand. In this context, scientists are increasingly being called upon to predict how physiological acclimation to increased future temperatures may affect behavior, growth and reproduction and possibly shape the long-term fate of species. For example, the aerobic scope of some tropical reef fishes is known to decrease when water temperature is raised above normal summer temperature, limiting their development and reproductive capacity (Rosa et al., 2012).

Temperature is known to be one the main factor regulating the duration of fish embryonic development. Oxygen consumption rates were significantly affected by temperature in different developmental stages. With elevated temperature, heat shock protein (Hsp) production may be induced several hundred fold to repair, refold, and eliminate damaged or denatured protein mechanisms. Additionally, the expected increase in metabolic demands with concomitant ROS formation may be followed by an enhancement in the activity of antioxidant enzymes (Lesser, 2006). The increase of heat shock protein level was indicated both in fish and in invertebrates at high temperature. A significant enhancement of the heat shock response (Hsp70/Hsc70) was shown in different developmental stages of squid while the GST and CAT activities varied unclearly with the exception of PER which increased significantly with temperature in early developmental stages (Rosa et al., 2012).

Embryonic development was dramatically shortened with a concomitant negative effect on growth and survival success. Even though the underlying causes of embryonic death outside the range of normal development are not known, decreased membrane permeability, disequilibria of coupled enzyme reactions, and limits imposed by kinetics and inactivation of enzyme proteins could be some of the responsible mechanisms likely to play a role. The increased metabolic load incurred by squid embryos resulted in loss of growth potential, and after hatching, paralarvae were less developed (premature) and showed greater incidence of abnormalities. The investigators suggest that the acceleration of the hatching process was mainly promoted by the higher oxygen demands coupled with hypoxic stress (see 5.1). In fact, premature hatching is a common response to hypoxia in fish (Kamler, 2008; Rosa et al., 2012).

Increased temperatures led to higher embryonic metabolic rates. To increase the flow of oxygen by means of diffusion, invertebrate eggs swell, leading to greater surface areas and reduced egg wall thicknesses (Cronin and Seymour, 2000). Yet, such swelling does not prevent pO_2 from consistently falling (to critical levels) and pCO_2 from rising within eggs as it was demonstrated at the case of cephalopod eggs (Gutowska and Melzner, 2009; Hu et al., 2010). The elevated catabolic activity seemed to have accelerated hypoxia within egg capsules, and caused metabolic suppression. This finding was corroborated with thermal sensitivity data, which revealed that late embryos displayed Q_{10} values below 1.5, indicative of active metabolic suppression (Grieshaber et al., 1994). Concomitantly, and regardless of temperature, the fermentative glucose-opine pathway was more used by embryos. This may be linked to the reduced capacity to extract oxygen at hypoxic and hypercapnic conditions within eggs (Hu et al., 2011). These stressful abiotic conditions inside eggs are expected to be aggravated under the projected near-future ocean warming, with deleterious effects on embryo's survival and growth (Rosa et al., 2012).

The energy loss due to swimming activity, which account for a large proportion of the total larvae energy budget, is expected to increase with warming. The authors investigated early development stages of squid, and they have shown that the "metabolic burst" associated with the transition from an encapsulated embryo to a jet propelled planktonic stage increased linearly with temperature. Therefore, the jet-propelled squid hatchlings will require more food per unit body size at higher temperatures and feeding failures will be critical since they show high metabolic rates and low levels of metabolic reserves. The time period for which new hatchlings can survive without food is known to be very limited at higher temperatures and so, they will need more food yet have less time to find it before facing mortality (Rosa et al., 2012). The authors also suggested that thermal tolerance range of embryos is lower than those found in larvae, because once larvae hatch, the temperature range over which they can persist increases significantly (Jordaan and Kling, 2003). Independent of the thermal environment, the embryo was more heat-tolerant. The lack of physical protection provided by the egg masses after hatching may contribute to the lower thermal tolerance of the hatchlings. Also, their lower thermal tolerance limits may be related to the high oxygen demands associated with the planktonic life strategy, which is coupled with an inefficient mode of locomotion (Rosa et al., 2012).

The impact of warming in the early life stages of fish also depends on the type of developmental modes within and among species. The greater exposure to environmental stress by the hatchlings seems to be compensated by physiological mechanisms that reduce negative effects of stress on fitness. The *Hsc70/Hsp70* concentrations were greatest in the near-future warming scenario for early developmental stages, with hatchlings showing the largest increase. The increased metabolic demands faced by the hatchlings must lead to elevated ROS formation, and *Hsps* are among the molecules that can eliminate/change the molecular configuration of ROS. Concomitantly, warming also led to an augment of TBA-compounds concentrations, indicative of the enhancement of lipid peroxidation. In the projected summer warming scenario, the increased oxygen requirements of the hatchlings led to higher SOD activity. This indicates that there was an increase of superoxide production. CAT activity also was enhanced possibly to increment the capability to catabolize peroxide resulting from SOD action. These biomarkers constituted an integrated stress response in larvae, but not in embryos, to ocean warming (Rosa et al., 2012).

The researchers characterized the response of the *Hsp70* protein family to heat-shock in several developmental stages of the freshwater fish medaka *(Oryzias latipes)* (Werner et al. 2004). At each of the five developmental stages, embryos ($n = 100$ per stage) were given a 30-minute heat-shock at 104°F and returned to culture conditions. Control embryos were maintained at 77°F. In early-stage embryos (stage 11, early gastrula), stress-protein (*Hsp70*) concentrations did not increase in response to heat-shock, while *Hsp70* proteins were induced in all other developmental stages. These early-stage embryos were considerably less tolerant to heat stress than embryos at stage 19 and older.

Of stage 11 embryos heat-shocked at 104°F for 30 minutes, 22% died within 1 day. By day 3 another 22% died. By day 4 another 28% showed retarded development as compared to controls, with the formation of large, clear spaces around the heart and over the yolk sac. In addition, eyes were absent or fragmented. These alterations have also been described in medaka embryos exposed to various toxicants, and represent pericardial edema, yolk sac and peritoneal edema, and microphthalmia (reduced growth of the eyes resulting in small eyes or blindness), respectively. Only 28% developed normally (Werner et al., 2004).

Hatching began after 7 days, and by day 14, 93% of normally developed embryos had hatched. Of the hatched larvae, 23% had curved spines. Stage 19 embryos heat-shocked at 104°F for 30 minutes developed normally, but initiated hatching earlier than the embryos heat-shocked at later stages and the control embryos. 80% hatched on day 7 of development whereas hatching had occurred in only 0% to 14% of all control embryo groups. By day 9, all embryos heat-shocked at stages 19 to 35 and control embryos hatched, and showed no sign of developmental lesions. However, a small percentage of stage 23 and stage 26 embryos showed "helmet heads," which occurs when all but the head emerges from the chorion and this structure remains closely attached to the head (Werner et al., 2004).

Therefore, the stress-protein response is one of the most important cellular mechanisms to prevent and repair the adverse effects of thermal stress, as well as of some chemical pollutants as described above.

5.3 UV-irradiation Effects

Light is a strong biological regulator of physiological processes in fish, such as feeding, digestion, and reproduction. Light has an influence on the signaling systems in the fish brain and the retina of the eye. It comprises different components such as quantity (light intensity), quality (spectral composition), light distribution (point source or evenly), and cycling (photoperiod and season). One of the negative consequences of climate change is associated with the depletion of ozone layer and the increase of UV-irradiation. Unfavorable light conditions influence negatively on larvae development, because they do not facilitate good feeding behavior or that induce stress reaction as in the case of high level of UVR. UVR carries more energy per photon than any other wavelength. Such highly energetic photons initiate the following negative processes in living organisms (Zagarese and Williamson, 2001):

- Firstly, they can generate reactive oxygen species (ROS), which cause oxidative stress to cells and tissues and damage DNA and proteins in living organisms.
- The second property is ubiquity: owing to their dependence on light, primary producers and most visual predators, such as fish, are also necessarily exposed to damaging levels of UVR.
- Thirdly, the combined effect of UVR and additional environmental factors may result in synergistic effects, such as the photoactivation of organic pollutants and photosensitization.

Aquatic organisms have evolved three basic mechanisms to cope up with UV-irradiation: vertical migration in the water column, photoprotection (e.g., production of photoprotective compounds such as antioxidants and specially adapted pigments), and dark repair mechanisms of DNA-damage sites (Steeger et al., 1999; Kohler et al., 2001). UVR impairs sperm motility, reduces fertilization, and causes embryo malformation that in turn affects recruitment and therefore the sustainability of natural populations. The direct molecular effects of UVR are mediated by absorption of certain wavelengths by specific macromolecules and the dissipation of the absorbed energy via photochemical reactions. Most organisms have defense mechanisms that either prevent UVR-induced damage, or mechanisms that repair the damage (Dahms and Lee, 2010; Browman et al., 2001). Both the egg and juvenile stages are transparent in color, with integument pigmentation becoming apparent much later in development. This lack of protective melanin in the integument used for the absorption and dissipation of high energy UV-B renders these earlier life stages more vulnerable to UV-B–mediated damage (Charron et al., 2000).

UVR impacts the living systems on various levels of their biological organization from molecular to ecosystem and biosphere (Fig. 5.1). At present this problem is very important because it plays a role in global changes of the environment associated which significantly damage the natural physical and chemical processes in the biosphere. Ozone layer depletion was documented for the first time in Antarctic region in 1985, and it has been increasing during the last 30 years, and could result into catastrophic consequences for life on the Earth. There have been exhaustive studies on the responses of terrestrial organisms on the UV-irradiation, while the information of aquatic organisms is very scarce, in spite of 70% of the planet being covered, water ecosystems produced one third of the total productivity of biosphere. It was estimated that the decrease on 25% of ozone layer the UV-irradiation upon the ocean surface will increase and phytoplankton photosynthesis will drop to 35%. This will lead to decrease of zooplankton and fish number. At the same time the sea food

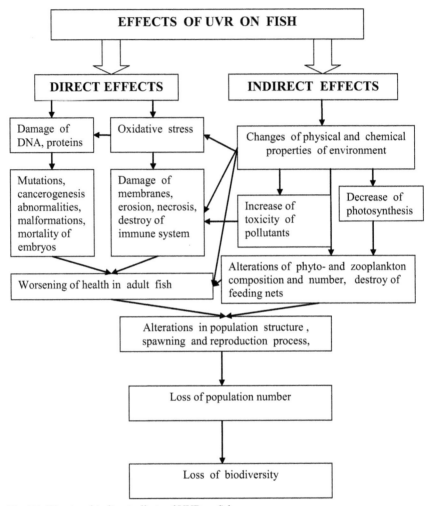

Fig. 5.1. Direct and indirect effects of UVR on fish.

covers approximately 18% of protein consumption of humans especially in developing countries, whose population grows gradually (Kohler et al., 2001).

The investigators documented that the various fish species adapt to habitats with high UV-B radiation (Fukunishi et al., 2006). For instance, embryos of *Danio rerio* derived from zebrafish raised in outdoor ponds were more tolerant to UVB than the embryos from laboratory-raised fish (Dong et al., 2007) and generally zebrafish embryos had greater tolerance to UVR and was a good model for detection of UV damage (Dong et al., 2008). The ultraviolet-B (UV-B) radiation (UVBR: 280–320 nm) may strongly impact planktonic fish larvae. The consequences of an increase of UVBR on the osmoregulatory function of *Dicentrarchus labrax* larvae were investigated. In young larvae of *D. labrax*, in which as in other teleosts, osmoregulation depends on tegumentary ion transporting cells, or ionocytes, mainly located on the skin of the trunk and of the yolk sac. As early *D. labrax* larvae passively drift in the top water column, ionocytes are exposed to solar radiation. A loss of osmoregulatory capability was shown in larvae after 2 days of low (50 µW cm^{-2}: 4 h L/20 h D) and medium (80 µW cm^{-2}: 4 h L/20 h D) UVBR exposure. Compared to control larvae kept in the darkness, a significant increase in blood osmolality, abnormal behavior and high mortalities were detected in larvae exposed to UVBR from day two onwards. At the cellular level, an important decrease in abundance of tegumentary ionocytes and mucous cells was observed after 2 days of exposure to UVBR. In the ionocytes, two major osmoeffectors were immunolocalized, the Na$^+$/K$^+$-ATPase and the Na$^+$/K$^+$/2Cl$^-$cotransporter. The authors hypothesize that the impaired osmoregulation in UVBR-exposed larvae originates from the lower number of tegumentary ionocytes and mucous cells. This alteration of the osmoregulatory function could negatively impact the survival of young larvae at the surface water exposed to UVBR (Sucré et al., 2012).

Embryos and larvae of zebrafish (*Danio rerio*) were exposed to UV to determine the effects on expression of CYP1 and stress response genes *in vivo* in these fish. Embryos were exposed to UV on two consecutive days, with exposure beginning at 24 and 48 h post-fertilization (hpf). Embryos exposed for 2, 4 or 6 h twice over 2 days to UVB (0.62 W/m^2; 8.9–26.7 kJ/m^2) plus UVA (2.05 W/m^2; 29.5–144.6 kJ/m^2) had moderately (2.4 ± 0.8-fold) but significantly up-regulated levels of CYP1A. UVA alone had no effect on CYP1A expression. Proliferating cellular nuclear antigen (PCNA) and Cu–Zn superoxide dismutase (SOD1) transcript levels were induced (2.1 ± 0.2 and 2.3 ± 0.5-fold, respectively) in embryos exposed to two 6-h pulses of 0.62 W/m^2 UVB (26.8 kJ/m^2). CYP1A was also induced in embryos exposed to higher intensity UVB (0.93 W/m^2) for two 3-h or two 4-h pulses (20.1 or 26.8 kJ/m^2). CYP1B1, SOD1 and PCNA expressions were induced by the two 3-h pulses of the higher intensity UVB, but not after two 4-h pulses of the higher intensity UVB, possibly due to impaired condition of surviving embryos, reflected in a mortality of 34% at that UVB dose. A single 8-h long exposure of zebrafish larvae (8 dpf) to UVB at 0.93 W/m^2 (26.8 kJ/m^2) significantly induced CYP1A and CYP1B1 expression. The authors conclude, that UVB can induce expression of CYP1 genes as well stress response genes in developing zebrafish, and that UVB intensity and duration influence the responses (Behrendt et al., 2010).

Direct impact of ambient (1.95 W/m^2) and subambient doses of UV-B radiation on muscle/skin tissue antioxidant status was assessed in mature zebrafish (*Brachydanio rerio*). The influence of these doses on hatching success and survival in earlier life stages was also examined. Subambient doses of UV-B radiation in the presence (1.28 W/m^2) and absence (1.72 W/m^2) significantly depressed muscle/skin total glutathione (TGSH) levels compared to controls (0.15 W/m^2) and low (0.19 W/m^2) UV-B–treated fish after 6 and 12 h cumulative exposure. Ambient UV-B exposure significantly decreased muscle/skin glutathione peroxidase (GPx) activity after a 6 h exposure; activities of glutathione reductase (GR) were unchanged. Superoxide dismutase (SOD) and catalase (CAT) activities peaked after 6 and 12 h cumulative exposure, respectively, but fell back to control levels by the end of the exposure period. The changes in tissue antioxidant status suggested UV-B–mediated increases in cytosolic superoxide anion radicals and hydrogen peroxide and in muscle/skin thiobarbituric acid reactive substances (TBARS). Hatching success of newly fertilized eggs continuously exposed to ambient UV-B was only 2% of the control value. Even at 30% and 50% of ambient UV-B, hatching success was only 80% and 20%, respectively, of the control. Newly hatched larvae exposed to an ambient dose of UV-B, experienced 100% mortality after a 12 h cumulative exposure period. This study supports a major impact of UV-B on both the mature and embryonic zebrafish (Charron et al., 2000).

Laboratory exposures of embryos and larvae of Atlantic cod (*Gadus morhua*) to ultraviolet radiation (UVR) equivalent to a depth of approximately 10 m in the Gulf of Maine resulted in significant mortality of developing embryos and a decrease in standard length at hatching for yolk-sac larvae. Larvae at the end of the experimental period also had lower concentrations of UVR-absorbing compounds and exhibited significantly greater damage to their DNA, higher activities and protein concentrations of the antioxidant enzyme SOD and significantly higher concentrations of the transcriptional activator p53. p53 is expressed in response to DNA damage and can result in cellular growth arrest in the G1- to S-phase of the cell cycle or to apoptosis. Cellular death caused by apoptosis is the most likely cause of mortality in embryos and larvae exposed to UVR, while the smaller size at hatching in those larvae that survived is caused by permanent cellular growth arrest in response to DNA damage. Sub-lethal energetic costs of repairing DNA damage or responding to oxidative stress may also contribute to poor individual performance in surviving larvae that could also lead to increase in mortality. Therefore, expression of the p53 pathway, may result in the smaller size at hatching observed for cod larvae exposed to UVR. Because the principal function of p53 is to promote survival or deletion of cells exposed to agents that cause DNA damage, such as hypoxia, UVR, ROS or mutagens, larvae that survive the effects of UVR at the biochemical and molecular levels, there must be an energetic cost at the organism level (Lesser et al., 2001; Lesser, 2006). A corresponding dose-dependent increase in the *Hsp70* protein levels was also observed in sea urchin embryos exposed to different UVB doses. The investigators detected variations in the *Hsp70* mRNA levels for any of the UV-B doses, and they proposed a post-transcriptional regulation of UV-B-induced stress response (Bonaventura et al., 2006; Wu et al., 2007).

Monitoring of pike larvae exposed to UV-B showed that the neurobehavioral syndrome revealed substantial late mortality. As UV-B had no influence on CYP1A content in larva, retene (9–82 µg l⁻¹) induced this protein substantially with and without UV-B. In pike, the applied UV-B radiation and water retene both decreased hsp70 concentrations. Neither UV nor retene changed SOD activity significantly. Overall, data on pike suggested that only a minor increase in ambient UV-B coming to the earth's surface may cause lethal effects to larval fish (Hakkinen et al., 2004).

The sub-lethal energetic costs of repairing DNA damage or responding to oxidative stress may contribute to poor individual growth performance of fish larvae. If the small, but significant, decrease in size at hatching observed in the experiments occurs in nature, it may result in a longer planktonic period for the fish and increase the probability of being predated upon. In addition to predicting mortality, modeling efforts on the effects of UVR on fish and invertebrate larvae should also include the energetic costs of sub-lethal effects and how these costs affect other life-history phases in addition to predicting direct mortality (Lesser et al., 2001). For this purpose we investigated the UV-irradiation effects on heat production of Black Sea fish *Atherina mochon pontica* larvae. Larvae (size 9.3–15.9 mm) were caught in the coastal waters of Sevastopol Bay and transferred in the glasses with marine water. Fish were exposed to UV-irradiation 1 min. Bactericide lamp Kvartz-240 was used as a source of UV-radiation (280–400 nm). The heat production determination was indicated in the Thermometric 2277 Thermal Activity Monitor (LKB, Sweden) at +20°C due 60 h. In control and experimental groups we used 5 individuals. The results obtained showed that the survival of exposed larvae was not significantly different as compared to control group. However the values of heat production of irradiated larvae was significantly lower than in untreated ones (Table 5.1).

Table 5.1. Heat production of *A. mochon pontica* larvae (mean ± SEM, n = 5), exposed to UV-irradiation for 1 min.

Values of heat production	Intact larvae	UV-affected larvae	P
Heat production, µWt/larvae	21.1 ± 1.7	11.8 ± 3.0	< 0.05
Heat production, µWt/mg wet weight	2.1 ± 0.7	1.2 ± 0.3	< 0.05
The highest value of heat production, µWt/larvae	26.3 ± 2.3	16.1 ± 4.4	n/s

n/s—not significant

The mean values of metabolic rate of exposed larvae were lower more than 2-fold than the values of intact fish. The obtained results demonstrated the significant damage of fish larvae metabolism under UV-irradiation that could negatively influence their growth, development and reproduction. We noted similar effect in *Artemia* larvae exposed to UV-irradiation (Shaida and Rudneva, 2004). Thus the increase of UV-radiation could result in great hazard of water ecosystems and the marine organisms especially in early life.

Thus, climate changes associated with the growth of eutrophication, temperature increase and UV-irradiation led significant damage of metabolic processes, connected with the imbalance of energetic mechanisms which resulted in the organism's death. Heat production and other biomarkers are good tools for global climate changes effects on biota and could be applied for evaluation of the negative biological consequences and their prediction. Moreover, many fish species experience large diurnal and seasonal temperature alterations which make them a favored subject for studying the mechanisms that underly environmentally induced physiological plasticity and adaptations to climate changes in various geographical locations. The irradiances of UVB radiation that elicit these responses in fish embryos and larvae can occur in many temperate latitudes, where these ecologically and commercially important fish are known to spawn, and may contribute to the high mortality of embryos and larvae in their natural environment. The application of biomarkers as "early warning system" of negative biological events could predict them and develop the remedies and protective challenges.

References

Almesjo, L. and H. Limen. 2009. Fish populations in Swedish waters How are they influenced by fishing, eutrophication and contaminants? The Riksdag Printing Office, Stockholm. 70 pp.

Baganz, D., G. Staaks, S. Pflugmacher and C.E.W. Steinberg. 2004. A comparative study of microcistin-LR induced behavioral changes of two fish species (*Danio rerio* and *Leucaspius delineatus*). Environ Toxicol. 19: 564–570.

Behrendt, L., M.E. Jönsson, J.V. Goldstone and J.J. Stegeman. 2010. Induction of cytochrome P450 1 genes and stress response genes in developing zebrafish exposed to ultraviolet radiation. Aquatic Toxicol. 98(1): 74–82.

Bieniarz, K., P. Epler, D. Kime, M. Sokolowska-Mikolajczyk, W. Popek and T. Mikolajczyk. 1996. Effects of N, N-dimethylnitrosamine (DMNA) on *in vitro* oocyte maturation and embryonic development of fertilized eggs of carp (*Cyprinus carpio* L.) kept in eutrophied ponds. J. Appl. Toxicol. 16: 153–156.

Bonaventura, R., V. Poma, R. Russo, F. Zito and V. Matranga. 2006. Effects of UV-B radiation on development and hsp70 expression in sea urchin cleavage embryos. Marine Biol. 149: 79–86.

Browman, H.I., R.D. Vetter, C.A. Rodriguez, J.J. Cullen, R.F. Davis, E. Lynn and J.-F. St. Pierre. 2001. Ultraviolet (280–400 nm)-induced DNA damage in the eggs and larvae of

Calanus finmarchicus G (Copepoda) and Atlantic Cod (*Gadus morpha*). J. Exp. Biol. 204(1): 157–164.

Borisova, O., A. Kondakov., S. Palear, E. Rautalahti-Miettinen, F. Stolberg and D. Daler. 2005. Eutrophication in the Black Sea region. Impact assessment and Causal chain analysis. University of Kalmar. Sweden. 60pp.

Charron, R.A., J.C. Fenwick, D.R.S. Lean and Th.W. Moon. 2000. Ultraviolet-B radiation effects on antioxidant status and survival in the zebrafish, *Brachydanio rerio*. Photochemistry and Photobiology. 72(3): 327–333.

Conley, D.J., C. Humborg, L. Rahm, O.P. Savchuk, F. Wulff. 2002. Hypoxia in the Baltic Sea and basin-scale changes in phosphorus biogeochemistry. Environ Sci Technol. 36; 5315–5320.

Cronin, E.R. and R.S. Seymour. 2000. Respiration of the eggs of the giant cuttlefish *Sepia apama*. Mar Biol. 136: 863–870.

Dahms, H.U. and J.-S. Lee. 2010. UV radiation in marine ectotherms: Molecular effects and responses. Aquatic Toxicol. 97(1): 3–14.

Deane, E.E. and N.Y.S. Woo. 2007. Impact of nitrite exposure on endocrine, osmoregulatory and cytoprotective functions in the marine teleost *Sparus sarba*. Aquatic Toxicol. 32: 85–93.

Deng, D.F., K. Zheng, F.-Ch. Teh, P.W. Lehman and T.J. Swee. 2010. Toxic threshold of dietary microcystin (-LR) for quart medaka. Toxicon. 55: 787–794.

Diaz, R.J. and R. Rosenberg. 1995. Marine benthic hypoxia: A review of its ecological effects and the behavioural responses of benthic macrofauna. Oceanogr. Mar. Biol. Annu. Rev. 33: 245–503.

Diaz, R.J. and R. Rosenberg. 2008. Spreading Dead Zones and Consequences for Marine Ecosystems. Science. 321: 926–929.

Dong, Q., K. Svoboda, R. Terrence, T.R. Tioersch and W.T. Monroe. 2007. Photobiological effects of UVA and UVB light in zebrafish embryos: evidence for a competent photorepair system. J. Photochem. Photobiol. 88(2-3): 137–146.

Dong, Q., W.T. Monroe, R. Terrence, T.R. Tioersch and K.R.J. Svoboda. 2008. UVA-induced recovery during early zebrafish embryogenesis. Photochem. Photobiol. 93(3): 162–171.

Edwards, M. and A.J. Richardson. 2004. Impact of climate change on marine pelagic phenology and trophic mismatch. Nature. 430: 881–884.

Engstrom-Ost, J., M. Lehtiniemi, S. Green, B. Kozlowsky-Suzuki and M. Viitasalo. 2002. Does cyanobacterial toxin accumulate in mysid shrimps and fishvia copepods? J. Exp. Mar. Biol. Ecol. 276: 95–107.

Engstrom-Ost, J., M. Karjalainen and M. Viitasalo. 2006. Feeding and refuge use by small fish in the presence of cyanobacteria blooms. Environ Biol Fish. 76. 109–117.

Engstrom-Ost, J. and U. Candolin. 2007. Human-induced water turbidity alters selecion on sexual displays in sticklebacks. Behavioral Ecology. 18: 383–398.

Engström-Öst, J., E. Immonen, U. Candolin and J. Mattila. 2007. The indirect effects of eutrophication on habitat choice and survival of fish larvae in the Baltic Sea. Mar. Biol. 151(1): 393–400.

Fukunishi, Y., R. Masuda and Y. Yamashita. 2006. Ontogeny of tolerance to and avoidance of ultraviolet radiation in red sea bream *Pagrus major* and black sea bream *Acanthopagrus schlegeli*. Fisheries Science. 72(2): 356–363.

Gazenave, J., M. Bistoni, E. Zwirnmann, D.A. Wunderlin and C. Wiegand. 2006. Attenuating effects of natural organic matter on microcystin toxicity in zebra fish (*Danio rerio*) embryos—benefits and costs of microcystin detoxication. Environ. Toxicol. 21(1): 22–32.

Gutowska, M.A. and F. Melzner. 2009. Abiotic conditions in cephalopod (*Sepia officinalis*) eggs: embryonic development at low pH and high pCO_2. Mar Biol. 156: 515–519.

Grieshaber, M.K., I. Hardewig, U. Kreutzer and H.O. Pörtner. 1994. Physiological and metabolic responses to hypoxia in invertebrates. Reviews Physiol. Biochem. Pharmacol. 125: 143–147.

Hakkinen, J., E. Vehniainen and A. Oikari. 2004. High sensitivity of northern pike larvae to UV-B but no UV-photoinduced toxicity of retene. Aquatic Toxicol. 66: 393–404.

Hassell, K.L., P.C. Coutin and D. Nugegoda. 2009. A novel approach to controlling dissolved oxygen levels in laboratory experiments. J. Exp. Mar. Biol. Ecol. 371: 147–154.

Hu, M.Y., E. Sucre, M. Charmantier-Daures, G. Charmantier and M. Lucassen. 2010. Localization of ion-regulatory epithelia in embryos and hatchlings of two cephalopods. Cell Tissue Res. 339: 571–583.

Hu, M.Y., T. Yung-Che, M. Stumpp, M.A. Gutowska and R. Kiko. 2011. Elevated seawater PCO$_2$ differentially affects branchial acid-base transporters over the course of development in the cephalopod *Sepia officinalis*. Am. J. Physiol. Reg. Integr. Comp. Physiol. 301. R1700–R1709.

Jordaan, A. and L.J. Kling. 2003. Determining the optimal temperature range for Atlantic cod (*Gadus morhua*) during early life. In: H.I. Browman and A.B. Skiftesvik (eds.). The Big Fish Bang Proceedings of the 26th Annual Larval Fish Conference. Bergen, Norway: Institute of Marine Research. pp. 45–62.

Kamler, E. 2008. Resource allocation in yolk-feeding fish. Rev. Fish Biol. Fish. 18: 143–200.

Kankaanpaa, H., P.J. Vuorinen, V. Sipia and M. Keinanen. 2002. Acute effects and bioaccumulation of nodularin in sea trout (*Salmo trutta m. trutta* L.) exposed orally to *Nodularia spumigena* under laboratory conditions. Aquat Toxicol. 61: 155–168.

Karjalainen, M., J. Engstrom-Ost, S. Korpinen, H. Peltonen, J.-P. Paakkonen, S. Ronkkonen, S. Suikkanen and M. Viitasalo. 2007. Ecosystem consequences of cyanobacteria in the northern Baltic Sea. Ambio. 36: 195–202.

Kohler, J., M. Schmit, H. Krumbeck, M. Kapfer, E. Litchman and P. Neale. 2001. Effects of UV on carbon assimilation of phytoplankton in a mixed water column. Aquatic Sciences. 63: 294–309.

Lesser, M.P. 2006. Oxidative stress in marine environments: biochemistry and physiological ecology. Ann. Rev. Physiol. 68: 253–278.

Lesser, M.P., J.H. Farrell and C.W. Walker. 2001. Oxidative stress and p53 expression in the larvae of Atlantic cod (*Gadus morhua*) exposed to ultraviolet (290–400·nm) radiation. J. Exp. Biol. 204: 157–164.

Levitus, S., J. Antonov and T. Boyer. 2005. Warming of the world ocean, 1955–2003. Geophys Res Lett. 32.

McElroy, A., C. Clark, T. Duffy, B. Cheng, J. Gondek, M. Fast, K. Cooper and L. White. 2012. Interactions between hypoxia and sewage-derived contaminants on gene expression in fish embryos. Aquat. Toxicol. 108: 60–69.

Meehl, G.A., T.F. Stocker, W.D. Collins, S. Solomon, D. Qin and M. Manning (eds.). 2007. Global climate projections. Climate Change 2007: The Physical Science Basis Cambridge: Cambridge University Press. 686–688.

Mustafa, S.A., S.N. Al-Subiai, S.J. Davies and A.N. Jha. 2011. Hypoxia-induced oxidative DNA damage links with higher level biological effects including specific growth rate in common carp, *Cyprinus carpio* L. Ecotoxicology. 20(6): 1455–1466.

Ojaveer, E., M. Simm, M. Balode, I. Purina and U. Suursaar. 2003. Effect of *Microcystis aeruginosa* and *Nodularia spumigena* on survival of *Eurytemora affinis* and the embryonic and larval development of the Baltic herring *Clupea harengus membras*. Environ. Toxicol. 18: 236–242.

Palikova, M., R. Krejai, K. Hilscherova, B. Buryskova, P. Babica, S. Navratil S., R. Kopp and L. Blaha. 2007. Effects of Different Oxygen Saturation on Activity of Complex Biomass and Aqueous Crude Extract of Cyanobacteria During Embryonal Development in Carp (*Cyprinus carpio* L.) Acta Vet. Brno. 76: 291–299.

Perry, A.L., P.J. Low, J.R. Ellis and J.D. Reynolds. 2005. Climate change and distribution shifts in marine fishes. Science. 308: 1912–1915.

Pörtner, H.O. and R. Knust. 2007. Climate change affects marine fishes through the oxygen limitation of thermal tolerance. Science. 315: 95–97.

Pörtner, H.O., F.C. Mark and C. Bock. 2004. Oxygen limited thermal tolerance in fish? Answers obtained by nuclear magnetic resonance techniques. Resp. Physiol Neurobiol. 141: 243–260.

Qiu, T., P. Xie, Z. Ke, L. Li and L. Guo. 2007. *In situ* studies on physiological and biochemical responses of four fishes with different trophic levels to toxic cyanobacterial blooms in a large Chinese lake. Toxicon. 3: 365–376.

Randall, D.J., C.Y. Hung and W.L Poon. 2004. Response of aquatic vertebrates to hypoxia. In: Fish Physiology, Toxicology, and Water Quality. Proceedings of the Eighth International Symposium, Chongqing, China, October 12–14, 2004. 1–11.

Rosa, R. and B.S. Seibel. 2008. Synergistic effects of climate-related variables suggest future physiological impairment in a top oceanic predator. Proc. Nat. Acad. Sci. USA. 105: 20776–20780.

Rosa, R., M.S. Pimentel, J. Boavida-Portugal, T. Teixeira, K. Trübenbach and M. Diniz. 2012. Ocean Warming Enhances Malformations, Premature Hatching, Metabolic Suppression and Oxidative Stress in the Early Life Stages of a Keystone Squid. PLoS One. 7(6): e38282.

Rubenchik, B.L. 1990. Cancerogenes Formation from the Nitrogen Compounds. Naukova Dumka. Kiev (*in Russian*).

Rudneva, I.I. and E. Petzold-Bradley. 2001. Environmental and security challenges in the Black Sea region. In: E. Petzold-Bradley, A. Carius and A. Vimce (eds.). Environment Conflicts: Implications for Theory and Practice. Netherlands: Kluwer Academic Publishers. pp. 189–202.

Rudneva, I.I., E.B. Melnikova, S.O. Omelchenko, I.N. Zalevskaya and G.V. Symchuk. 2008. Seasonal variations of nitrosamine content in some Black Sea fish species Turkish J. Fisheries and Aquatic Sciences. 8: 283–287.

Schlenk, D. and R. Lavado. 2011. Impacts of climate change on hypersaline conditions of estuaries and xenobiotic toxicity Aquatic Toxicol. 105 (3–4): 78–82.

Shaida, V.G. and I.I. Rudneva. 2004. The effect of UV-radiation on brine shrimp Artemia *Artemia salina* L. Proceedings of Odessa National University. 9(5): 133–139.

Steeger, H.-U., M. Wiemer, J.F. Freitag and R.J. Paul. 1999. Vitality of plaice embryos (*Pleuronrvtes platessa*) at moderate UV-B exposure.) J. Sea Research. 42(1): 27–34.

Strand, J., L. Andersen, L. Dahllöf and B. Korsgaard. 2004. Impaired larval development in broods of eelpout (*Zoarces viviparus*) in Danish coastal waters. Fish Physiol. Biochem. 30(1): 37–46.

Sucré, E., F. Vidussi, B. Mostajir, G. Charmantier and C. Lorin-Nebel. 2012. Impact of ultraviolet-B radiation on planktonic fish larvae: Alterations of the osmoregulatory function. Aquatic Toxicol. 109: 194–201.

Thomas, P. and M.S. Rahman. 2009. Biomarkers of hypoxia exposure and reproductive function in Atlantic croaker: A review with some preliminary findings from the northern Gulf of Mexico hypoxic zone. J. Exp. Mar. Biol. Ecol. 381 (Supplement 1). S38–S50.

Thomas, C.D., A. Cameron, R.E. Green, M. Bakkenes and L.J. Beaumont. 2004. Extinction risk from climate change. Nature. 427: 145–148.

Werner, I., S.L. Clark and D.E. Hinton. 2004. Biomarkers aid understanding of aquatic organism responses to environmental stressor. California Agriculture. 57(4): 110–115.

Wiegand, C., N. Meems, M. Timoveyev, C.E.W. Steinberg and S. Pflugmacher. 2004. More evidence for humic substances acting as biogeochemicals on organisms. In: E. Chubbour, and G. Davis (eds.). Humic Substances: Nature's Most Versatile Materials, New York: Taulor & Francis. pp. 349–362.

Winston, G.W., C.A. Traynor, B.S. Shane and A.K.D. Hajos. 1992. Modulation of the mutagenicity of three dinitropyrene isomers in vitro by rat-liver S9, cytosolic, and microsomal fractions following chronic ethanol ingestion. Mutation Research. 279: 289–298.

Wu, R.S.S, E.W.H. Shang and B.S. Zhou. 2004. Endocrine disrupting and teratogenic effects of hypoxia on fish, and their ecological implications. In: Fish Physiology, Toxicology, and Water Quality. Proceedings of the Eighth International Symposium, Chongqing, China, October 12–14, 2004. 75-87.

Wu, Y., D. Xing, L. Liu, T. Chen and W.R. Chen. 2007. Fluorescence resonance energy transfer analysis of bid activation in living cells during ultraviolet-induced apoptosis. Biochem. Biophys. Acta. 39(1): 37–45.

Zagarese, H.E. and C.E. Williamson. 2001. The implications of solar UV radiation exposure for fish and fisheries. Fish and Fisheries. 2(3): 250–260.

Conclusions and Perspectives

The results of many investigations obtained recently documented that fish embryos and larvae are used to evaluate the interaction among common environmental (chemical, physical and biologicalo) stressors found in aquatic environments, namely xenobiotics, hypoxia/anoxia, salinity, increasing temperature, UV-irradiation, microalgae toxins, pathogens, etc. Fish embryos and larvae are more sensitive to these factors as compared to adults and at the molecular level, the early life stages responding to these stressors share common response factors, and evidence exists for cross-talk between them. The range of endpoints used on the fish embryo tests allows an integrated analysis that contributes to a better understanding of the toxicity and mode of action of different kinds of toxicants and physical stressors in experimental and field conditions. Although many studies have been devoted to the effects of pollutants on various physiological and biochemical parameters in fish embryos and larvae, in many cases their precise mechanism of action remains to be clarified.

In the case of field studies more complex environmental mixtures of toxicants include compounds that may act through many different direct and indirect mechanisms and pathways and interact with each other. Their effects may be synergistic and antagonistic and may cause different consequences for fish development and growth. Little is known about how such cocktail effects affect these fish responses. Thus, effects of multiple stressors, evolutionary toxicology, and the relative roles of technical and conceptual limitations to our understanding of chemical and physical effects on aquatic systems play an important role in future development of the methods and tools of environmental risk assessment and water quality evaluation. Hence, the greatest challenge in environmental toxicology is to understand effects of mixture toxicity not only in the case of chemicals, but with physical factors also. Chemical permissible levels of pollutants are traditionally based on experiences from laboratory experimental studies with single chemicals. However, chemicals can be more toxic when mixed with other chemicals, or impact by physical factors (UVR), because the cocktail effect is a more realistic scenario at field conditions.

Future works could focus on investigation of modes of action on early life stages of fish because embryo and larvae testing were revealed to be so informative, a refinement of the test could be made including other endpoints (toxicogenomics) for the effect of exposure to different kinds of chemicals and physical factors. Recently, ecotoxicogenomics is developing into a key and informative tool for the assessment of environmental impacts and environmental risk assessment for aquatic ecosystems because gene expression is a sensitive indicator of toxicant effects at the cellular level. For biomarkers to fulfill their potential, they should be mechanistically relevant and reproducible. In the perspective of risk assessment, these endpoints should be explored in order to assess their usefulness as early warning system and correlations should be sought between short-term tests and effects of long-term exposures as it is observed in more realistic scenarios in environment.

Biomarker approach has been widely and convincingly used to study the mechanism of action of toxicants and to identify the organism response for the identification of toxic contaminants. These mechanisms are just beginning to be revealed in marine organisms, especially in adults, but their characterization will be fundamental for better understanding of the implications of their development and variations in early fish stages including gene expressions according to fish embryogenesis and larvae.

More knowledge is needed of the physical factors (temperature, light, hypoxia/anoxia, salinity, UV-radiation) effects on fish physiology and biochemistry through early ontogenesis, because these factors are essential for hatching success and larval metabolism and growth. Potential benefits of better understanding the effects of temperature, light, salinity, and oxygen on larval fish performance may be significant especially in optimization of aquaculture conditions. Understanding the physiology of hypoxia/anoxia and oxygen level acclimatization in larvae of different species may prevent conditions that can lead to incidents of fish losses in intensive culture and in environment.

In addition, the effects of wind, waves, noise and sun radiation, which are attributed to the open sea surface, on fish embryos and larval development and performance are unknown. The specific biomarkers of these effects on fish eggs and larvae are needed. Thus, the analysis of biomarker level in different fish species, in different stages of development taking into account their phylogenic position, specific features of ecology and physiology is important for understanding the key ways of evolution of life, evaluation of fish abilities to protect against pollutants and keep their life and biodiversity in the impact environments caused stress and disturb animals health. Thus, more biomarkers are needed for understanding the possible effects on embryos, larvae and juvenile fish.

Index

Color Plate Section

Chapter 2

Fig. 2.6. Black Sea Scorpion fish (*Scorpaena porcus*).

Chapter 3

Fig. 3.1. Developing embryos and adult fish of round goby *N. melanostomus*.

Chapter 4

Fig. 4.18. *Atherina hepsetus* adult and its larvae.

Printed and bound by CPI Group (UK) Ltd, Croydon, CR0 4YY

23/10/2024

01778227-0009